Condition Monitoring of Mechanical and Hydraulic Plant

Condition Monitoring of Mechanical and Hydraulic Plant

A concise introduction and guide

Trevor M. Hunt

*Independent Consultant, formerly of Rolls-Royce Ltd
and the Bath Fluid Power Centre*

CHAPMAN & HALL

London · Glasgow · Weinheim · New York · Tokyo · Melbourne · Madras

Published by Chapman & Hall, 2–6 Boundary Row, London SE1 8HN, UK

Chapman & Hall, 2–6 Boundary Row, London SE1 8HN, UK

Blackie Academic & Professional, Wester Cleddens Road, Bishopbriggs, Glasgow G64 2NZ, UK

Chapman & Hall GmbH, Pappelallee 3, 69469 Weinheim, Germany

Chapman & Hall USA, 115 Fifth Avenue, New York, NY 10003, USA

Chapman & Hall Japan, ITP-Japan, Kyowa Building, 3F, 2-2-1 Hirakawacho, Chiyoda-ku, Tokyo 102, Japan

Chapman & Hall Australia, 102 Dodds Street, South Melbourne, Victoria 3205, Australia

Chapman & Hall India, R. Seshadri, 32 Second Main Road, CIT East, Madras 600 035, India

First edition 1996

© 1996 Trevor M. Hunt

Typeset in 10/12pt Palatino by Mews Photosetting, Beckenham, Kent

Printed in Great Britain by St Edmundsbury Press Ltd, Suffolk

ISBN 0 412 70780 2

A catalogue record for this book is available from the British Library

Library of Congress Catalog Card Number: 96-83038

∞ Printed on permanent acid-free text paper, manufactured in accordance with ANSI/NISO Z39.48-1992 and ANSI/NISO Z39.48-1984 (Permanence of Paper).

Contents

Preface

'It is finished!' is possibly the greatest phrase ever spoken on earth. To start is good, but to bring to satisfactory completion is what really matters. This is the subject of this book.

The beginning of the 20th century saw a tremendous rise in the production of mechanical machinery, followed by hydraulic equipment as the century progressed. However, it soon became apparent that machinery did not last as long as one would like, and a new science developed called 'maintenance' to help keep the equipment on the go for a longer time. Maintenance is a large and complex subject; indeed the few comments in the first chapter of this book reveal that there are over 20 different types of maintenance. Gradually, however, maintenance workers discovered that to be truly effective there was a need of a further feature, and that feature is 'condition monitoring'.

Stop for a moment and consider this progression:

a machine starts … it needs to continue running
… it completes its task

The way it is designed and made and built, enables it to 'start'. The way it is operated enables it to 'run'. However, the ancillaries, the supplies, the connections to the outside world, the environment, even going back to the design and operation, these all affect its 'life' – whether it will reach satisfactory completion.

This is where maintenance and condition monitoring come in, and they come in, hand in hand. Maintenance is designed to provide the necessary service input from the outside world so that life continues. Condition monitoring ensures that the maintenance is done as mechanically effectively and cost effectively as possible. Maintenance without condition monitoring is like taking medication every day for life without checking whether it is still needed or even whether it was needed in the first place. Condition monitoring without maintenance is like knowing you are sick but not bothering to go to the doctor.

This book, however, is primarily devoted to the condition monitoring aspect. It covers a very wide range of ideas and techniques. It is a practical book, and one which is written to encourage all involved in machinery and fluid power to look around at what can be done. There is sufficient detail to appreciate the finer points of each type of monitoring, sufficient to start along the best route for the application in mind.

Part One is the background to condition monitoring. It is important to be in the correct frame of mind when approaching the subject, and this part draws out the importance of the whole concept. It introduces various approaches to the subject, in particular the 'CME' approach.

Part Two is the heart of the matter. Here are all the monitors, and what a range there is! It is somewhat concentrated owing to the restricted length of the book, but this part will be valuable as a reference long after the book has first been read.

Part Three sees the monitoring in use. There are warnings, too, because not all attempts at monitoring have been successful, and we need to watch out for the pitfalls.

Part Four is the reference section as regards tables and further reading and definitions. Most ideas are defined in the text of the book, but a list of common initials is given each with its basic meaning. Finally, there are many names and addresses of companies who have given information for the book, either directly or indirectly from their literature; they would be delighted to help the reader further in their quest for effective condition monitoring.

We may not be able to say 'It is finished!' with quite the same eternal significance as the original speaker, but hopefully, through this book, we will be able to achieve a more than satisfactory life for our equipment in a manner which is cost effective.

Trevor Hunt
Bristol
June 1995

PART ONE

The Approach to Monitoring

CHAPTER 1

Take any machine

This is a book about the monitoring of a machine's condition. There are many monitors for evaluating the machine's product, but they do not normally give guidance regarding the machine condition – only how it has been set up. On the other hand, if the setting is correct and an unacceptable product is beginning to develop, then the output may well be one way of assessing machine condition. However, there are many more, much clearer and earlier, means of detecting deterioration.

This opening part is designed to enable us to appreciate what condition monitoring is and how it is essential for consistent and reliable machine operation.

1.1 THE PURPOSE OF MACHINERY

A 'machine' is a mechanism for change. It manipulates material. It modifies movement. It manages human resources. In each case the original form of something has been changed and, hopefully, brought to a new improved condition. A machine, therefore, usually has a clear objective, and its purpose is to achieve that objective, to the best of its ability, within the constraints placed on humans of time, space and matter. In other words

- in the right period   –   time,
- in the right place   –   space,
- in the right condition   –   matter.

Consider the four simple examples in Table 1.1. It is apparent from the table that not everything is as precise as one would like. The words 'as possible' carry a certain feeling of variability which may not be acceptable in the ultimate machine; instead there could be a definite value stated, such as, for example 2 in Table 1.1.A, flight BA501 will take 2 hours 35 minutes and will land at Heathrow airport and will arrive with the words on the intercom 'We hope you enjoyed the flight'.

Table 1.1 Examples of the purpose of 'machines'

Example	Machine	Time	Space	Matter
1 Machine a valve block from a solid block of steel	Milling machine etc.	As quickly as possible	With dimensions as close to the drawing as possible	Heat treated and surfaced to give the best life in service
2 Carry 200 people from A to B	Aeroplane	As quickly as possible	To arrive as close to the destination as possible	Everyone well, and pleased with the flight
3 Extract oil from a North Sea stratum	Oil rig	As quickly as possible	Oil brought out and stored in a container.	The oil retained in as pure a condition as possible
4 Prevent flood waters entering an important region	Hydraulically actuated flood gates	As quickly as possible	To seal totally as required	Remain strong enough to hold all the water

Whatever the machine we have, or the system we use, we should be very clear as to its purpose. Without that, the rest of this book will be pointless. Even this book has a purpose – to help you to achieve the satisfactory fulfilment of the purpose of your machine–system. It cannot always be done, but by using an appropriate technique the purpose can be considered as something achievable – with confidence.

Whether the purpose needs to be considered in its entirety, or as a specific feature, will vary depending on the system. For instance, a plant will have the total purpose of producing a certain product to a certain quality in a certain time. That output can be monitored, and any deviation from the required level can be signalled. However, individual items within the plant will each have their own specific purposes. This can be broken down even further into the smaller components within the individual items.

It is quite feasible for a faulty component to be so easy to replace, and for this to be done so quickly, that to monitor for a fault could actually cause a delay rather than speed up the process. In effect, the part of the system is so simple, although important, that the need for individual monitoring is not necessary. One example would be the replacement of spark plugs on an internal combustion engine; here the overall engine will be monitored in some way but not the specific components.

Occasionally the purpose of the 'machine' is such that monitoring is totally unwarranted. One could think of the target aircraft used for artillery practice; here the need is for low cost with no concern for engine life.

1.2 REASONS FOR THE PURPOSE NOT BEING ACHIEVED

We all know that not all civil aircraft land on time. Some land early. Some do not even take off. It is not surprising, therefore, that you will find in the small print of an airline schedule something like 'Every effort is made to ensure punctuality of our services, but we cannot accept liability for the consequences of a delay, or a cancellation of any flight and cannot guarantee making connections.' (Major airlines even run an overbooking arrangement expecting that some travellers will not turn up, but if they do all arrive at the airport not all will be able to travel.)

So why is there this lack of absolute certainty?

This applies to not just aircraft, but to all machinery. Also, people come into the equation as well, as seen in the flight comments above. In a world which is constantly aware of decay, where mistakes are made by human error, where Acts of God are beyond our control, we have little chance of being 100% reliable. The company which advertised its product in a recent trade journal with such confidence (Fig 1.1) is either one which makes nothing or one which is totally naive.

Figure 1.1 100% reliable?

Within the book various examples will be given of faults developing and how they can be detected at an early stage, but just consider how common they are. Table 1.2 gives a few of the more clear-cut examples.

The cause of failure – lack of fulfilment of the purpose – may be just one feature or a combination of many. It may, for instance, be a combination of human error, machine wear and outside intervention. For the

Table 1.2 Examples of cause of failure

Industry	Machine	Fault which could lead to failure
Automobile manufacture	Robot welding machine (valve)	Orifice blockage
Aerospace	Aero engine (lubricator)	Oil leak–starvation
Quarrying	Earth mover (lever arm)	Actuator wear
Marine	Diesel engine (injector)	Nozzle blockage
Mining	Drive engine (gear box)	Tooth fracture
Manufacturing	Hydraulic press (pump)	Cross-port cavitation

moment it is essential to grasp that there will be failure and it can come from a variety of sources.

Apportionment of blame is difficult, as when a fork-lift truck operator was trying to save his employer money by coasting down a long ramp with the hydraulic pump switched off. The problem was the operator did not know that the truck was power steered, and when he reached the bottom of the slope he needed to turn right just in front of a brick wall. The life of the truck, the goods it was carrying, and the brick wall, were all curtailed, but whose fault was it? Could it have been avoided by monitoring? Perhaps a more comprehensive 'fool-proof' design would have helped, but also, we could say, comprehensive monitoring could have detected such a hazard developing (e.g. wheel motion when there was no power).

A major failure reason, if not the major failure reason, is the human operator. Inadequate training, forgetfulness, illness, mental instability, a prior bad experience causing thoughts to wander, alternative attractions, hunger, hate, love, tiredness, heart attack – all these can contribute to the effectiveness or otherwise of a machine operator. The result of that ineffectiveness can be a machine disaster. (Incidentally, while it may not be easy to monitor the person, the immediate machine deviation from the norm can be monitored, e.g. the dead-man's handle on a train.)

1.3 THE PURPOSE OF MAINTENANCE

Maintenance and monitoring go hand in hand.

As more maintenance is included, less monitoring is required. On the other hand, as more monitoring is included, less maintenance is required. It is the optimizing of maintenance–monitoring which achieves the lowest machine costs (see Fig. 5.2).

There is a wide range of possibilities, going from the sublime to the ridiculous. In other words, with a high level of maintenance, with a considerable cost in time and replacement parts, almost nothing is going to fail and there is little point in doing any monitoring. The problem is that the downtime will be enormous and the overall costs totally unacceptable.

On the other hand, if every conceivable item on the machine is monitored, then the exact moment for the replacement of a part may be determined, and the maintenance team cut to a handful to be called in at the most convenient time. Here there is little cost in maintenance, but the cost of monitoring will be totally out of proportion to the value of the machine.

Maintenance of machinery is a subject all of its own, but it is worthwhile here just to discuss briefly the possibilities between the two extremes mentioned above. There are many different types of maintenance and different people view them in various ways. However, in this context the chart in Fig. 1.2 may clarify the situation. The idea of the chart is to provide a sense of three features of machine maintenance.

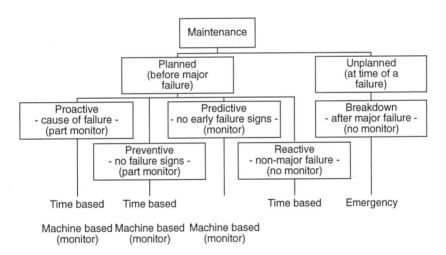

Figure 1.2 Maintenance types related to monitoring.

1. When is maintenance undertaken?
2. What does the maintenance do?
3. Is condition monitoring included?

It will be noticed how important the condition monitoring is in the critical maintenance types, i.e. the condition monitoring is the key to preventing failure. Each of the types (and many others) is detailed in the following listing (in alphabetical order after the overall definition). It must be understood, however, that the terminology is not as exact in the literature as it could be; for instance, 'corrective' is used in two quite different ways – 'putting right from predictive maintenance' and 'reactive maintenance'. There is also considerable overlap between the types.

Maintenance

This is the examination of the plant functions to assess their current efficient operation, the replacement of all features which look as if they are, or will be soon, deteriorating, the topping up of levels which have 'gone down' and generally bringing to the 'as-new' condition. It includes not only replacing and repairing but also servicing (bringing up to date with the latest modifications).

Breakdown maintenance

(This is sometimes called failure maintenance.) This is the maintenance which is actioned after the system or plant or component has failed. It is the unplanned maintenance which is costly and unattractive to personnel who are called on at the most inconvenient of times.

Condition-based maintenance

(See proactive, preventive and predictive maintenance.)

Corrective maintenance

This is based mainly on performance monitoring. Where performance is dropping, a certain amount of maintenance, in the sense of 'retuning', is undertaken to ensure that the best condition is maintained. It is expected and should be planned for. It may require changing with experience. It is not shown in Fig. 1.2, but could be considered as a combination of breakdown (unplanned) and reactive.

Failure maintenance

(See breakdown maintenance.)

Fixed-time maintenance

(See reactive maintenance and time-based maintenance.)

Improvement maintenance

This is that part of maintenance which involves modifications and redesign.

Machine-based maintenance

This term relates to maintenance before any breakdown. It may, however, be before any deterioration (such as proactive maintenance) or after the early stages of failure become detectable (such as preventive or predictive).

On-condition maintenance

This is a combination of proactive, preventive and predictive maintenance, as indicated by the machine monitoring.

Planned (or schedule) maintenance

This is where maintenance is considered as essential. In some way maintenance is scheduled to occur during the operation of the plant. A maintenance manager and team are appointed and generally they will operate at the most convenient time (convenient for them and convenient for the plant). Quite how this is done is defined by the possibilities shown in Fig. 1.2.

Predictive maintenance

By being able to predict the future deterioration of machinery from what has already happened it is able to determine the most convenient time for maintenance to be undertaken. Either the machinery may be examined at frequent intervals or a continuous monitoring may be undertaken.

Preventive (or preventative) maintenance

From an examination of a deteriorating current situation, maintenance is undertaken to prevent any further worsening. Normally such maintenance requires a certain amount of monitoring to provide the evidence to action the procedure at the best time; it can, however, be time based.

Proactive maintenance

From the monitoring of features which may cause failure – rather than the early signs of failure – suitable maintenance can be actioned to prevent any commencement of deterioration. More detail on proactive maintenance is given in section 4.1.3.

Reactive maintenance

This is the replacement of items which have already failed, i.e. the plant is examined from time to time (possibly on a regular basis), and items which have failed are replaced. This can only be acceptable for non-critical components such as lights.

Reliability-centred maintenance

This is an all-inclusive term for the systematic procurement of a long-term acceptability of the machinery, including reliability, safety and cost-effectiveness. It could be considered as machine-based maintenance covering all planned maintenance but with the examination of the complete failure pattern of complex machinery. It is based on the inherent reliability of each item of equipment when used in its correct operating state.

Time-based maintenance

(This is sometimes called fixed-time maintenance.) From past experience, and from careful analysis or guesswork if the system is new, a schedule of maintenance is organized so that at certain fixed intervals each part of the system is appropriately maintained. In order to be sure that no actual breakdown is experienced (predictive maintenance), the time interval will of necessity be shorter than it need be. In combination with breakdown maintenance it is possible to determine an optimum where some costs are due to corrective work and the other costs are from time-based action. However, it can only be acceptable in those cases where the mechanism of deterioration is related to time – such as wear. The fixed intervals may be either 'machine hours' or 'calendar hours'.

Total productive maintenance

This maintenance idea is designed to provide a continuous improvement in production. Normally production drops with time, owing to deterioration of the machinery and lack of personal interest. The objective of this maintenance is to promote interest and involvement from a variety of personnel in group activity. Maintenance is thus taken more seriously and will include routine inspection, cleaning and equipment problem solving in a genuinely co-operative manner.

Unplanned (or unscheduled) maintenance

It is assumed that the plant will not break down provided that it is operated correctly. However, should the plant stop as a result of some fault, then an emergency team will be hurriedly formed and asked to put the matter right immediately, however inconvenient it may be. It is run on emergency lines.

(There are also a number of more specific types of 'maintenance', such as 'protective maintenance' – which looks at surface coatings.)

Maintenance is thus seen as an essential part of any plant or system which is needed to last. The way it is done, the type of maintenance, however, has a strategic effect on the actual life. It is interesting to note that in 1992 one survey indicated that while the average percentage of cost of maintenance with respect to plant value was 4.2%, the better companies had been able to reduce this to the order of 2% – by careful planned maintenance including condition monitoring: 'Improvements in plant reliability and plant condition monitoring were naturally the key to upgrading production efficiency' (Hernu, 1992).

What type of maintenance is incorporated will depend on the flair of the management. It is easy to plan nothing (unplanned maintenance) and to become involved in crippling expense. It is relatively easy to set up a time interval for examination (time-based maintenance) and to have no failures but to see the financial resources dropping. It is not that much harder, though, to involve appropriate monitoring (machine-based maintenance) and to see the reliability of the machinery improving and unnecessary costs dwindling.

A complete programme of maintenance for a site or plant has considerable attraction. Total productive maintenance (TPM) and reliability-centred maintenance (RCM) come to mind. One suggestion has been that TPM and RCM are complementary, the first being directed by the plant operations team and the second being run by the engineering team; in other words, the involvement of both would provide the optimum situation.

Training courses are run on TPM and RCM; many books are also available and contract maintenance is on offer. Individual predictive maintenance computer programs are sold which are linked directly to condition monitoring. These are user friendly and enable a machine operator to be constantly aware of any changes developing in their equipment.

It should also be mentioned that numerous maintenance management programs can be purchased for machine systems and plant. Some of these are well-documented programs which can be used by any company. They are termed computer-aided maintenance (CAM) or computer-aided maintenance management (CAMM) or computerized maintenance management system (CMMS). They should cover such features as

- plant inventory,
- stock management,
- plant availability statistics with a service history,
- on-line downtime monitoring,
- reliability-centred maintenance with fault analysis and
- interfaces to proprietary condition monitoring packages.

This book is not a maintenance manual, but it has included the above remarks because of the close connection between maintenance and monitoring in well-maintained systems. However, care should be exercised before becoming involved in one particular maintenance schedule; many are related solely to one type of monitoring and this may or may not be the best for our system or plant. Hence it is necessary to be clear about the monitoring, before considering maintenance.

1.4 THE PURPOSE OF MONITORING

Appropriate monitoring minimizes overall expenditure. This has been discussed in part in the previous section – in relation to maintenance. The objective of condition monitoring is to prolong effective machine life. However, wrongly used it can not only reduce machine life, but also increase overall costs and hence the need to be aware of what is available, what is appropriate and what will bring the required saving.

Perhaps four terms could highlight the purpose of monitoring:

1. reliability;
2. life;
3. consistency;
4. cost saving.

1.4.1 Reliability

Reliability has been defined by three major organizations as follows:

- BSI   –   the characteristic of an item
  expressed by the probability[4]
  that it will perform[2] a required function[1]
  under stated conditions[3]
  for a stated period of time[5].
- EOQC – the measure of the ability of a product
  to function[1] successfully[2]

when required
for the period required[5]
in the specified environment[3].
It is expressed as a probability[4].

- NASA – the probability[4] of
a device performing[1] adequately[2]
for the period of time[5] intended
under the operating conditions[3] encountered.

Each of these includes five specific features:

1. function;
2. performance;
3. operating range;
4. probability;
5. life.

The purpose of monitoring is thus to achieve the required reliability for the system, plant or component. But in evaluating what reliability exists make sure that all the above five features are included; reliability without including all five is inadequate and meaningless.

1.4.2 Life

Life is really part of the reliability equation. To monitor the machine as regards its life is to ensure that the required life will be achieved (for the conditions etc. stated in the reliability description). If the monitoring is unable to hold the machine to this life, then it is not good enough and an alternative type of monitoring should be considered.

Life, of course, does not mean just ticking-over. Life must contain the whole purpose of existence if true satisfaction is to be gained – '*Life to the full*', as the Bible puts it. We need to expect a fully functioning machine with acceptable efficiency and output during the whole of its required life.

1.4.3 Consistency

A certain variety is permissible. Tolerances are normally stated for engineering components and output. However, if these limits are exceeded then the product is no longer acceptable. As a machine ages, its wear becomes more severe and consistency is unachievable; the monitor needs to be able to check on either the prime cause of wear or the wear itself. In some production applications it may be possible to monitor the output for size or whatever parameter is critical.

The consistency is then achieved by the maintenance process replacing any wearing parts at the most convenient time with the minimum of cost; in other words, bringing the machine back to its 'life' condition.

1.4.4 Cost saving

Ultimately, most processes are run in order to make a profit. They may be of value to humanity, but unless they make a profit, or at least are seen as 'good value for money', they cannot continue. Monitoring provides the advantage over systems which are run on a breakdown maintenance system. Monitoring may require a certain cost to start and a small cost to run, but it should be able to keep the total cost down, and certainly lower than if it had not been used.

As mentioned in the discussion earlier there may have to be a balancing of maintenance costs and monitoring costs in order to achieve the optimum.

1.5 SORRY ABOUT THE PEOPLE

This book only deals with machine monitoring. That means that one of the major causes of failure and downtime – the person operating the machine – is not discussed in detail. It should not, however, be disregarded; here are just two examples to start the thought process. A monitor such as a head-mounted 'drowsiness' detector (looking at the eyelids) may have a place in motorway driving, but it also could be invaluable with complex machine operators. Another monitor of great value in hazardous and security regions is the Erwin Sick proximity laser scanner which is able to detect the presence of the wrong person at a machine.

It is not easy to monitor men and women, because of their complexity and the unique freewill which is designed within. Fortunately, machines are more predictable.

CHAPTER 2

But do not take
any monitor

So, maybe it would be a good idea to monitor the machinery. Does anything come to mind? Well, there are some very highly advertised products – almost household names – perhaps they would be the thing, or perhaps they would not. It is one thing to sense a need, but it is another to ensure that the need is properly monitored. You would not use a stethoscope for an ingrowing toe-nail.

This chapter looks at the importance of relevance. We look at how a consistent route can be taken to ensure that the best monitor is fitted. It is a very simple route, but highly effective. It can be remembered by the initials **CME** (for mechanical engineers this is particularly easy to remember, especially for the Chartered Mechanical Engineer):

- **C**   stands for **component** – the parts that fail;
- **M**   stands for the **monitor** – how the failure will be detected;
- **E**   stands for **economics** – is it really going to be cost effective to monitor?

A typical example of the *CME* analysis process will be given in Chapter 11, but this chapter details the depth of meaning of each of the three features.

2.1 C – COMPONENT FAILURE ANALYSIS

First criticize the plant, the system, the components. Have a brainstorming session. Whether we like it or not, our system will fail. However, where will it fail? The more devious a mind is and the more experienced the engineer, the more outlandish will be the suggestions. Hence it is good to have a 'brainstorming' session to try to sense all conceivable possibilities (Table 2.1).

Before the brainstorming session, it could be helpful for some members who will be attending to consider the whole concept of 'failure'.

1. What constitutes a failure?
2. What functions cause a failure?

This is an enormous field, but if it is restricted to mechanical and hydraulic machinery, then definite guidelines are feasible.

Table 2.1 Brainstorming session

Present
   Machine operators and users
   Older experienced engineers
   Younger high flyers
      and designers.
Technique
   Explain the system or plant or component (or ensure all have had an opportunity to understand how it works and is fabricated)
   On a clearly visible board or sheet write down suggestions made as to where failures could occur – without any comment or criticism. Each idea will activate further thought from those present, until a very comprehensive list will be displayed.
Discuss the list
   Now discuss. Remove any unreasonable suggestions. What remains could be it.

2.2.1 Failure

A failure is where a function ceases to operate as required.

The 'function' could possibly be a negative function, i.e. something is expected to stop, but it continues to operate. In essence, the 'failure' is where the function behaviour is untimely and unwanted and is likely to lead to a loss of necessary output and an increase in cost.

2.2.2 Cause of failure

Failure is caused when the input to the function, the function itself and/or the output are adversely affected in some way.

The 'cause' can be that associated directly with the function, or it can be further back in the machine cycle or even the prime cause going back to the whole concept of the machine and its maintenance. These two concepts, the failure and the cause of failure, are best understood from a couple of practical examples. Figs. 2.1 and 2.2 show typical (although somewhat simplified) hydraulic and mechanical systems.

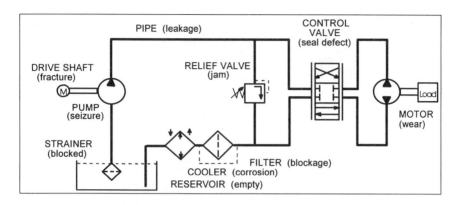

Figure 2.1 Typical failures in a hydraulic system.

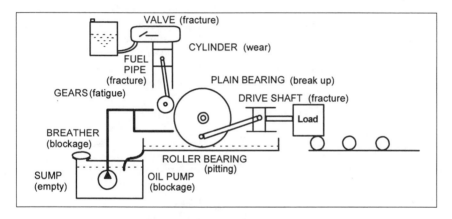

Figure 2.2 Typical failures in a mechanical system.

Marked on them, in parentheses, are a series of possible failures. Possible causes of these failures are shown in Table 2.2.

The examples in the table are merely to show a range of ideas, all of which can be critical in real systems. Maintenance will reduce the likelihood of these occurrences (as shown in the previous chapter) but to be able to monitor for them will give the added assurance that all is well.

After the brainstorming and listing of possible failures, it is best then to put some sort of weighting on the list. Put the list in order of criticality. Then mark the list with an order of likelihood of occurrence. Begin to sense what is important.

Chapter 3 extends the discussion on the components.

Table 2.2 Causes of failures

System	Failure	Possible causes
Hydraulic (Fig. 2.1)	Strainer blocked	Too fine a mesh
		Excessive large debris not cleaned out
	Pump seized	Oil starvation (upstream blockage)
		Orifice blocked (contamination)
		Shock load
		Seal failure
	Drive shaft fracture	Overload (pump seized)
		External blow
		Overheating (lack of lubricant)
	Pipe leakage	Pipe fracture (vibration)
		Poor coupling
		External damage
	Valve jammed	Debris too large (contaminant)
		Solenoid failure (burn out)
	Valve seal leaking	Wrong type
		Badly fitted
		Inadequate lubricant
	Motor worn	Excessive contaminant
		Lack of lubricant
	Filter blocked	Not changed when indicated
	Cooler corroded	Air–water in system
		Unsuitable materials
	Tank empty	Leakage (possibly elsewhere)
		Blockage (large debris ingested)
		Inadequate servicing–refilling
Mechanical (Fig. 2.2)	Oil pump blockage	Large debris
	Roller bearing pitting	Contaminant in the oil
		Shock loads
	Plain bearing break up	Excessive temperatures
		Too high a rotational speed
	Gear wear and fracture	Overload
		Lack of lubrication
		Misalignment
	Fuel pipe fatigued	Vibration
		Lack of clamping
	Valve fracture	Lack of lubricant
		High temperature
	Cylinder wear	Wrong lubricant
		Excessive speeds
	Breather blockage	Inadequate servicing–replacement
	Drive breakage	Gradual wear
		Overload
	Sump leakage	Bolt loose

2.2 M – OPTIMUM MONITORING

This book contains many examples of monitoring (Part Two gives the details), but this section is written to encourage the reader to be more selective in choosing the monitor. There are two concepts which need to be taken into account – the place of monitoring and the type of monitoring.

2.2.1 Place to monitor

The monitoring position needs to be relevant. It needs to provide the maximum evidence of what is happening in the system, to the plant, to the component. It may be an exceptionally efficient monitor, but misplaced it not only will be totally useless but may be completely misleading, giving a false security.

In oil and debris monitoring it is important to ensure that the debris is fully mixed in the oil or fluid being monitored; therefore positions must be chosen where this is most likely. Turbulent flow positions are best, e.g. immediately after a bend or connection, or just after a mechanical function such as a valve. Turbulence can be determined from the Reynolds' number (Table 2.3).

Table 2.3 Turbulent flow

Turbulent flow in a pipe occurs where the Reynolds' number (R_e) is greater than 4000, calculated from

$$R_e = vd/\nu$$

where v (m s^{-1}) is the liquid velocity, d (m) is the pipe internal diameter and ν (m^{2}s^{-1}) is the kinematic viscosity.

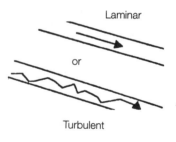

Laminar

or

Turbulent

Example
$v = 1\,\text{m s}^{-1}$, $d = 20\,\text{mm} \equiv 0.02\text{m}$, $\nu = 10\,\text{cSt} \equiv 0.000\,01\,\text{m}^2\text{s}^{-1}$, so $R_e = 2000$, which is still laminar.

A flow mixer could be installed in that part of the fluid circuit which otherwise exhibits laminar flow (Fig. 2.3). Sampling could also be where a free flow of oil enters a reservoir or sump (in this case it must be remembered that air will also be present and hence no monitor which is susceptible to air could be used). Further details on sampling are given in Chapter 9.

Figure 2.3 The Koflo pipe flow mixer insert (Pump & Package Ltd).

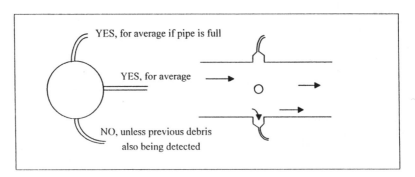

Figure 2.4 Best positions for sampling oil from a horizontal pipe for debris monitoring.

Another key question to answer concerns a horizontal pipe. Should the sampler be at the top, the side or the bottom of the pipe? It will depend on the requirement. If you need to know the slightest indication of debris in an otherwise well-cleaned fluid, then the bottom position may be best. If you want to know the average fully mixed contamination level then the top is possible, but, if air is likely to enter because the pipe is not full, it may be better to sample from the side (Fig. 2.4). (The reasoning here is that debris settles when the system is switched off.) For suggestions of the best positions in a hydraulic system for monitoring, see Fig. 7.23.

In vibration monitoring the detection direction will be important. Indeed, with suitable care the positioning can subtly filter out the vibrations which are not wanted (from adjacent machinery or other functions). The connection of an accelerometer (for vibration detection) needs to be as positive as possible; if the means of adhesion is poor then the higher frequencies will not be detected. (A screw-on sensor will give a much greater range of frequency, if required.)

Other monitoring types can be assessed in similar ways. Is the thermographic image going to show the part which is likely to fail? Is the corrosion sensor on the component which is likely to corrode?

The monitoring position needs to be convenient. No one will bother to fit a monitor if it cannot be fitted with ease and be detectable with ease. This means that not only the machine but also where that machine is finally to rest should be considered. If it is to go in a mine or a submarine, for instance, then greater care must be taken to ensure that the monitoring sampling point is where it can be easily reached. (One large hydraulic pump, known to the author, had two sampling points for different parts of the system; one sampler was on one side of the pump, the other on the other side. The problem was the pump was to be used down a mine in a restricted place, and there was no way both points could work – at least one of them was useless.)

In some cases it will be possible to fit a monitor at the time of build, but remember that it will still need to be serviced, so its position will still be critical.

2.2.2 Type of monitor

How relevant the monitor is requires a good knowledge of the failure mode – the reason for failure and the result of failure (i.e. the first indications rather than the final condition).

If the reason is fatigue, then some sort of frequency-based monitor is required. If the reason is contaminant based, then debris monitoring may be more suitable. There are many other questions which could be asked, each leading towards a different type of monitoring. Various examples are shown in Table 2.4.

Table 2.4 Some relevant monitors

Reason for failure	Signs of failure	Features to be monitored	Suitable monitor
Fatigue	Debris particles Rough running	Wear debris New frequencies	Debris Vibration Shock
Blockage	Reduced performance	Performance Steady state features, e.g. pressure, flow, temperature, speed	Performance Output
Corrosion	Debris particles Reduced thickness	Debris Potential etc.	Debris Corrosion
Overload	Temperature rise	Temperature Performance	Thermography Performance

Table 2.5 Monitoring features to be considered

Monitoring feature	Comments
1 Able to monitor all faults	Not very likely.
2 Easy to fit	Unless it is easy to fit, it will possibly never leave its box.
3 Easy to use	Imagine using it many times; is it still easy?
4 More reliable than the system being tested	The supplier will not admit to unreliability. Check with other users.
5 Able to be checked and calibrated	How often need this to be done? What is the cost of calibration or checking?
6 Able to give an immediate answer	How soon is an answer really required?
7 Able to be compensated for variables	Variables such as people, atmospherics, machine conditions.
8 Adaptable to different systems	Can it be used in other applications? (A possible cost saving.)
9 Able to store evidence for later checking	Always beneficial in any dispute or if further analysis required.
10 Inexpensive	In comparison with the overall cost of the machine operation. (Include the running cost as well as purchase price.)
11 Small and light	What size is acceptable? What size is preferable?
12 Non-iatrogenic	Will not cause a different sort of failure because it has been fitted
13 Safe	Electrically, chemically, mechanically and hydraulically.
14 Economically viable	Relates to 10 above. See section 2.3.
15 Easy to obtain and maintain	Applies to spares as well as the original purchase

In determining the best monitor there are many features to consider, but again the importance of each will depend on the application. Table 2.5 lists 15 features worth considering (with a brief explanation of each). No one monitor will be able to match one's exact need, but re-listing these 15 in order of priority will help to identify the most appropriate.

Chapter 4 looks in more detail at this optimization by considering the machine function and which part of that function could best be monitored.

2.3 E – COST EFFECTIVENESS

The cost of a monitor is always high, at first, but then there dawns on the mind the whole concept of the monitoring, which is to protect a machine from a possible high cost failure. We may begrudge paying a few pounds sterling for some medicine (it could always be cheaper) but it is nothing in comparison with the cost of losing our life. Was it not Queen Elizabeth I who offered her whole kingdom if someone could give her another year to live?

However, we have to be sensible. Is it really worth spending £100,000 on an analyser which will be used just a few times, when the work could be done by a laboratory in the same town for only £20 a go? If, for security reasons, it is essential that it be done in-house, then that is different; but there are few who have that reason.

On the other hand, to send data away, in a bottle or on a tape recording, makes no sense if it is going to take 3 days before a reply comes, and the plant has to remain idle during that time.

A subtle cost balancing is needed, as shown in Fig. 2.5.

Figure 2.5 The balance of costs.

These costs need to be calculated. It is cost effectiveness which is under discussion. Does the monitor make sense? Sadly, it is more common for a monitor to be fitted after a failure has occurred. It will not prevent the first horse leaving the stable, but the stable door bolted will prevent others leaving.

The cost equation is discussed in much greater detail in Chapter 5.

CHAPTER 3

The approach to component failure analysis

In order to evaluate fully the possible failure picture of a plant or system, it is essential to realize what happens in practice with the components which make up the machinery. This can be assessed in part by recalling all the failures experienced in the past. If they directly relate to the machine or system in question then that is a valuable step, but it is not everything. Every day new failures are occurring and we need to be more flexible in approaching the subject.

Quite frequently the cause of a failure is a multi-function. It may involve machine build, system design and operation, environmental conditions and the state of the man or woman who actually controls the machine. With so many variables, it is not surprising that dangerous combinations occur from time to time. In this chapter we look at the distribution of possibilities in order to be able to be prepared in our minds when approaching a failure analysis of our own system.

3.1 LATERAL THINKING

'Lateral thinking' is a plea to think of the less obvious. It is not easy. Not everyone can do this. However, those who can, after the immediate amusement generated, bring a wealth of suggestion to those present. In a way, the suggestions made may be preposterous, but, after the visionary, comes the realist who makes the suggestion into a reality.

In considering machine monitoring, the lateral thinker needs to know something about machines, but not too much. For instance, if a petrol pump is not working in a car, a person who merely know what a car does, rather than the intricacies of the pump itself, can come up with the following suggestions (in order from 'no lateral thinking' to 'maximum lateral thinking').

1. It has broken. Throw it away and replace it.
2. The mechanical–electrical activator of the pump is broken. Check.

3. There is no petrol entering the pump. Check the input line.
4. No signal is getting to the activator. Is the unit switched on?
5. The outlet is blocked. Check.
6. The bonnet catch is rough and as it is closed it disconnects the electrical supply.
7. There is a spider in the control which shorts out the circuit when it is warm.

Some of these suggestions may seem a little far fetched, but, in fact, I have met every one of them as reasons for the cause of failure of similar electrical or mechanical systems – including the spider.

How lateral you go is left to your imagination. However, basically the ideas which come should be encouraged; they will serve to illustrate the possibility of failure over a large range of conditions.

Although lateral thinkers will come up with the less obvious, one should also hold on to the obvious reasons as well. It is important that as wide a range of reasons as possible are considered in failure analysis. The following sections look at the mechanically more reasonable possibilities.

3.2 MACHINES DIFFER

Because one machine fails under certain conditions does not mean that all machines will similarly fail. Each machine is different; even if the

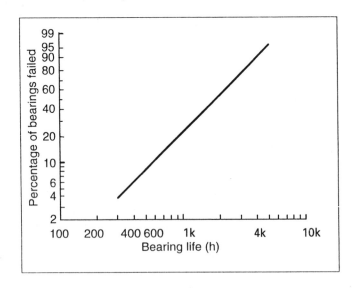

Figure 3.1 Statistical failure distribution for bearings.

design is identical, the manufacture and material will vary to a lesser or greater degree. If there are sufficient machines available, then a statistical analysis of failure could be undertaken – within the same range of conditions – and the likelihood of failure predicted. However, in the UK, apart from the well-known roller bearing tests (Fig.3.1), there is usually insufficient money to undertake such testing. Instead it is left to the engineer to 'estimate–guess' the probability of failure.

Conversely, of course, because one machine does not fail does not mean that all under similar conditions will also last. There is a measure of considerable uncertainty, particularly with each new design.

So what does one do? The best arrangement is to increase gradually the levels of a possible reason for failure until a failure does occur; at least that does give one result. However, whereabouts it is in the statistical distribution will not be known. Bayes' theorem (see a book on statistics) may help here.

Further testing could be undertaken with slightly different features, preferably to test for a failure induced for a different reason. Each such failure will indicate the mode of failure which will later have to be matched to the different types of monitors.

Another approach to the subject would be to sum the various machine tolerances, particularly the extremes, to see what change in temperature, friction, load, etc. may occur when the combination is at its worst. Each of these has the capability of activating a failure mode if not at the optimum dimension.

We have already looked at examples of the purpose of machines (Table 1.1). The examples in Table 3.1 illustrate a small range of different features in different processes which can be monitored to assess the

Table 3.1 Some special industrial features

Industry	Machine which might fail	Special features
Motor	Welding robot	Hydraulic Multi-axis Electrical
Process plant	Refinery pump	Chemical Pressure Sealing
Aero-engine	Blade forge	Temperature Material Load
Steel	Billet casting	Temperature Insulation Flow

machine health from their speciality. Anyone deeply involved in an industry will know those features which are special. Because they are special they could well be the most critical and the most likely to fail.

3.3 SYSTEMS DIFFER

This book does not look at electrical or chemical systems. If either of those is our subject of interest then perhaps some of the ideas can be cross-correlated. Here we are concerned with hydraulic and mechanical systems. However, neither are they the same as one another, nor are they the same throughout.

3.3.1 Hydraulic versus mechanical

It is true that both the mechanical system and the hydraulic system use fluids. Because of the fluid any vibration present will tend to be damped to a certain degree (hydraulic more so than mechanical). In some designs, the temperature transference (or cooling) can bring confusion to thermographic analysis.

However, it is in their different use of fluids that the greatest deviation arises. A hydraulic system uses the fluid – the hydraulic fluid – primarily as a means of power transference. The mechanical system uses the fluid – the lubricant – primarily as a means of reducing friction. (There are other secondary reasons such as cooling, damping, anticorrosion or biocidal.) The fluid is thus operating in totally different circumstances – the pressure exerted in hydraulic system pipes, which may reach many thousands of atmospheres, is totally different from that in lubrication system pipework, where the purpose is solely that of pushing the oil around to the next location.

Contaminant in fluids operating at very high pressures (hydraulic) can cause severe damage, as well as blocking the small refined orifices which are used. In a low pressure (lubrication) system contaminant might be considered at first sight as much more acceptable than that permissible in hydraulics. However, this is not necessarily true. The contaminant in lubricating oils may be just as disastrous. Much research has been undertaken here by bearing companies, and they are convinced that their bearings last considerably longer if the fluid is cleaner; even particles as small as $10 \, \mu m$ can cause severe wear.

So, maybe, there is not too great a difference between the hydraulic system and the mechanical system, but there are some deviations which it is important to note. A small hydraulic system is shown in Fig. 2.1. Similarly, a mechanical system is shown in Fig. 2.2.

3.3.2 Hydraulic system differences

Hydraulic systems have been designated 'high pressure' or 'low pressure', 'miniature' or 'standard', 'oil' or 'water', but the truth is that there is a wide and continuous range between each of these extremes. Even the oil–water mixtures vary in percentage. However, we certainly need to know the pressure and fluid which is being used.

The type of pump–motor used will vary and will have different modes of failure. For instance, if a piston pump design with a slipper pad still has the small lubricant hole within it, this orifice can block with disastrous effects. On the other hand, gear pumps have to cut themselves in and this debris is expected and is essential rather than being a disaster signal.

The types of valves used, both in the pumps and in the controllers, will vary considerably. A poppet valve, for instance, is much more prone to resist contamination than is a spool valve and may not need the same amount of monitoring.

What sort of service is the hydraulic system expected to give? Is it a mobile machine, or is it a fixed-bed stable system? Fig. 3.2 shows a couple of examples of hydraulic machinery which have completely different applications and need completely different monitoring.

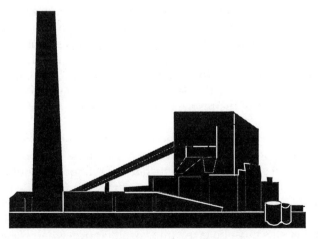

Figure 3.2 Two different types of hydraulic applications.

3.3.3 Mechanical system differences

Mechanical systems differ primarily in the drive, e.g. gears (spur, helical, spiral etc.), belts, shafts etc., in the bearings, e.g. rollers, balls, plain etc. The pump used will differ, as will the sump. How the system is cooled will vary. Speeds, reciprocating or rotating, cover a vast range.

It may not be immediately obvious why these differences are stated here. However, we need to sense the effect of these features on monitoring. For instance, a helical gear is designed to give a smooth action over a range of load, albeit with a sideways thrust. Fit a twin helical (herringbone) and the sideways thrust is cancelled. Spur gears are designed to give smooth action at one load only, and, as the load deviates from this optimum (either more or less), the vibration and wear will increase. Belts whip and fray. Plain bearings are soft; rolling element bearings are hard. Reciprocating motion may produce wear not too dissimilar to that of the rotating motion action, but it is a non-continuous sinusoidal effect and so vibration monitoring will differ.

The composite picture in Fig. 2.2 gives some further ideas.

3.4 ENVIRONMENTS DIFFER

Some environmental effects could be listed as shown in Table 3.2. In certain circumstances any one of these could cause a failure of a machine. It just depends how well it is isolated from the environment to which it is susceptible.

However, there are other 'environmental' features to be taken into account. There are the close-at-hand (and some not quite so close) other machines which are generating influences. Such effects as sound and vibration can be transmitted through the ground and cause considerable problems to adjacent machines. One example was the out of balance of a rotating machine which caused severe brinelling of the bearings in another machine a couple of metres away, but on the same foundational structure.

The statistical likelihood of some of these occurrences will vary from place to place (country to country). Where there is a quite low possibility, but human life is at stake, then even such low likelihoods should be taken into account. If health and safety are not part of the function, then it is more reasonable to consider only the most likely effects and to take out an insurance against the others.

Environmental effects must be taken into consideration when considering component failure analysis.

Table 3.2 Failure effects from different environments

Environmental condition	'Low' effect	'High' effect
Temperature	Reduced flow of fluids Frozen controls	Oxidation of fluids Burning Heat treatment
Humidity and water	(Dry)	(Wet) Corrosion
Light	Visibility errors on controls	Dazzling Reflection Fading
Earthquake	NA	No limit
Storm (e.g. wind, lightning, rain and hail)	NA	Breakage Seizure
Gases	(Low leakage) Accumulator discharge Pilot line control failure	Corrosion Human failure
Disease	NA	Human failure

Insects and birds	NA	Damage
		Circuit shorting or
		  blocking
		Removal of protective
		  surfaces
		Human disturbance

Mammals and reptiles	NA	Damage
Fish	NA	Blockage
Icebergs	NA	Damage
Plants	NA	Corrosion Blockage

3.5 OPERATORS DIFFER

Each human has his or her own special idiosyncracies. The way a clutch is pressed or a lever is moved, the visual assessment of a situation or a sight line. To make matters even worse, the pattern varies depending on the mental condition of the person; if they have just had a row, missed the bus, tripped over the cat, arrived later and been ticked off, their judgement will be severely warped.

Because of this, and the critical human–machine interface, machine failures are more prevalent than they need to be.

Any examination of a machine for failure modes must include the human factor, if it is possible to assess it. Occasionally a machine can be in 'auto-pilot' where the operator is put aside completely, but there is still the possibility of the operator overriding the auto-pilot and causing havoc. We have all experienced a traffic warden trying to order the traffic at a busy crossing, where the automatic time control has gone down; we find the queues are far longer.

A particular dangerous situation exists where only one person has been trained to operate a machine. There are written instructions, but if the trained, skilled, worker is off sick the likelihood of failure may quadruple if a stand-in takes over.

One way of assessing the human influence is to examine the instructions given to potential 'machine' operators. In what ways can those instructions be misread or confused? We need to try with a range of skilled or semiskilled or unskilled personnel – see what they make of the wording in the booklet and the icons on display and what happens if someone who will not even read the instructions takes on the job.

3.6 CHECK LIST

Perhaps the details given in this chapter may present a formidable list of hazards, enough to put anyone off. However, we should not despair, if

Table 3.3 Checking for possibilities of failure

Have I checked?		Suggestions	Possibility rating
Differences in component	1. 2. 3. 4.		
Differences in machines	1. 2. 3. 4.		
Differences in systems	1. 2. 3. 4.		
Differences in environment	1. 2. 3. 4.		
Differences in personnel	1. 2. 3. 4.		

Possibility ratings: 1, not very likely; 2, just a slight possibility; 3, a possibility; 4, 50:50 possibility; 5, more than just likely; 6, likely; 7, highly likely; 8, very likely indeed.

we have saved just one expensive 'unlikely' failure, we have probably paid for our wages, if not many others' too.

We need to have a check list as in Table 3.3 and to decide which plant, which systems and which components may fail.

CHAPTER 4

The approach to optimum monitoring

We have already considered possible positions for monitors to be attached to a machine (in Chapter 2), but the optimum position is something extra. Where on the machine is the best position? While that may be immediately obvious in some cases, the word 'obvious' is a signal for disaster. It prevents further discussion, when to the enquiring mind there are other options. It is preferable to stop and ask a more general question first.

Is the key failure mode signal best sensed on the input to the machine, on the machine itself, or on the output from the machine?

While it is understood that the failure we are discussing is occurring on the machine, or some part of it, it may actually be what is entering the machine which causes the failure. If that can be monitored, then the whole monitoring procedure is an extra step ahead of failure. Conversely, the fault developing may so influence the output (or the input) that if that is monitored then a quicker clearer picture of the failure development may be achieved more easily.

Each of three concepts will now be discussed in detail in order to develop an awareness of what is possible.

4.1 INPUT TO THE MACHINE

Looking at what goes into a machine can be rewarding in two aspects, possibly three:

1. we may sense something foreign in the input, which may lead to a failure,
2. we may be able to sense a drop in the acceptance of the input which conveys a machine failure and, possibly,
3. we may be able to stop the commencement of failure altogether.

4.1.1 Detecting something foreign entering

We have already considered what may cause a failure (e.g. Table 1.2). Most of the examples given were internal causes, and hence would not be monitored by an input sensor. However, there are also other possibilities, which are external; consider those given in Table 4.1. These are all very standard features, but each can be just the critical factor which causes the machine to stop.

Table 4.1 Outside causes of machine failure

Cause	Connection	Effect	Machine type
Atmospheric contamination (e.g. quarry dust)	Through filler cap, seals, air breathers	Blockage, jamming, wear	Fluid power and lubricated bearings
Vibration (e.g. from adjacent machinery)	Foundation, ground, supports	Fatigue (particularly if resonant frequency)	Rotating systems
Environment (e.g. humidity, temperature etc.)	Direct	Corrosion, locking	All
Low input conditions (e.g. flow)	Pipe work from reservoir, for example	Low pressure on input – cavitation	Hydraulic
High input conditions (e.g. temperature)	Pipe work from reservoir, for example	Oxidation and acidic attack	All
Operator behaviour	Controls	Erratic action	All except automatic control

4.1.2 Detecting a drop in the acceptable input

The obvious example here is power. If a fault has begun to develop, then the consumption of power, for the same output, will rise, e.g. electric power for an electric motor which is connected to a pressure-compensating pump. If the machine has failed completely, then the power may go down, rather than up, owing to the free running of the system. Either way, there will be a change in the consumption.

Another example could be the flow of fluid entering a process pump. This will be changed if a fault, such as leakage, has developed in the system.

4.1.3 Stopping a failure commencing

This is the proactive monitoring or proactive maintenance first described in Chapter 1 (section 1.3). This is by far the most valuable of the monitoring techniques if it is achievable. It means that no part of the machine needs replacing, but rather a better housekeeping or maintenance programme should be installed. The objective is to contain the suggestions made in Table 4.1 and to stop them influencing the machine by suitable maintenance at an early stage.

Probably the worst offender in Table 4.1 is contaminant in hydraulic and lubricant oils. This is a major subject on its own and one on which many books have been written, e.g. my own handbook (Hunt, 1993). However, the following highlights can be mentioned here.

- Contaminant may be liquid, gaseous or solid.
- Contaminant may be built in, absorbed, or generated.
- Contaminant is believed to cause over 70% of all fluid power failures.
- Contaminant in oil at a size as low as 5 μm can seriously reduce the life of bearings.
- Cleanliness in the oil should be maintained at an acceptable level.

The way components are manufactured and assembled can cause problems if inadequate cleaning and flushing are not actioned at the build stage. Corners and small 'dead ends' can retain solid and liquid contaminant if the cleaning process is only cursory: that is, it will retain it until some slightly higher than usual flow pulsations occur in the system and it is dislodged.

The level of cleanliness required should be stated by the supplier of equipment and should be maintained by the user. (Recommended levels and the ISO 4406 cleanliness chart are given in Appendix A.)

An example of life reduction in bearings due to the presence of water in the oil can be seen from Fig. 4.1.

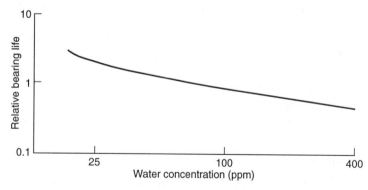

Figure 4.1 Reduction in bearing life due to water in oil.

Chapter 6 will go into detail about the types of monitors which are available for monitoring the input to machinery.

4.2 THE MACHINE ITSELF

Monitoring the machine itself is no surprise. Probably that is what we felt this book is all about. However, it needs to be tested against the possibility of monitoring the input to, or monitoring the output from, the machine.

Consider the possible advantages of the 'on-machine' monitor:

1. close to the failure mechanism, with less confusion from other features;
2. close to the failure mechanism, with better signal-to-noise ratio;
3. immediate response.

4.2.1 Less confusion from other features

In a large machine shop, or, indeed, a small one, there may be a variety of machines and operations taking place at the same time. Also, even more to the point each one is likely to be at a totally different point in any failure mechanism which may be developing. Thus, unless the sensor used (or sample taken) is close to the one machine being monitored the result may well be a confusion of several signals–machines. The closer the sensor is to the offending component or part, the clearer will be the signal.

On-machine monitoring has a distinct advantage here.

4.2.2 Better signal-to-noise ratio

This is somewhat similar to the comments just made above. However, it is really to highlight the sensitivity of a signal. 'Noise' need not be sound which we hear. 'Noise' can be electrical or some other radiation, which in some way can influence a sensor unless it is shielded, or unless it is too remote. The advantage of the on-machine monitor is that the 'noise' will be a minimum.

4.2.3 Immediate response

With rapid signal communication (by wire or wireless, infra-red or light, etc.) this may not be so appropriate. However, with certain sensors (particularly oil or those involving fluid transfer) the signal may be delayed depending on the oil line used. Again, on-machine will be the best with the shortest transmission lines.

It is very important to sense in this preliminary discussion on the approach to optimum monitoring that each type of failure in each type of situation will have its own specifications for the optimum monitor. We need to be reminded of the actual failures which are likely to occur and to relate them to what monitoring is possible.

Chapter 7 is a very detailed chapter with many types of monitors suitable for on-machine monitoring.

4.3 OUTPUT FROM THE MACHINE

Two valuable items come out of the machine, as regards machine condition monitoring:

1. the product – being manufactured, or service supplied;
2. the by-product – the effluent.

Either one of these may contain within it signals of distress due to a failure beginning to occur.

4.3.1 The product

Consider first the purpose of the machine – the product. In a sense this is the performance of the machine in practical terms. Such 'product' examples would be

- power,
- speed and
- parts manufactured.

It may be possible to arrange for a genuine 'performance monitor' to be built, but this will have to bring together the input consumption and the output produced – in other words an efficiency monitor. (This is discussed further in Chapter 8.)

4.3.2 The by-product

The unwanted discharge from a machine, the by-product, can consist of a variety of different features such as

1. metal swarf or material off-cuts left over in the machining,
2. contaminated or used oil,
3. noise,
4. gas and smell, and
5. heat.

Monitoring these may be difficult unless considerable experience has been built up. There is bound to be some 'waste', but how much has it to increase before it becomes a signal of inefficiency?

4.4 CHECK LIST

Here is a simple check list (Table 4.2) to see what possibilities for monitoring there might be. It will not convey how to monitor – that will be discussed later – but it will enable us to sense what items could be monitored.

Table 4.2 Checking for possibilities of monitoring

Have I checked?		*Suggestions*	*Possibility rating*
What is entering the machine?	1. 2. 3. 4.		
What is happening on the machine?	1. 2. 3. 4.		
What is coming out of the machine?	1. 2. 3. 4.		

Possibility ratings: 1, not very likely; 2, just a slight possibility; 3, a possibility; 4, 50:50 possibility; 5, more than just likely; 6, likely; 7, highly likely; 8, very likely indeed.

CHAPTER 5

The approach to cost effectiveness

Condition monitoring must be cost effective. Too frequently in the past an idea has been incorporated, at no mean cost, only to prove marginal, to say the least. Today, the manager is more conscious of financial benefits than of fascinations. No costs associated with future possibilities can be justified easily, hence the need to know – for sure – that what is being fitted will work and will save money.

This chapter puts into perspective the whole concept of small expense **now** to save large expense **later**. It may be true. It may not be true. Hence, it is important that we are truly realistic in our assessment of costs for each particular application. The total costs are what really matters – the 'terotechnology' concept – and this will be outlined first, followed by discussion on whether the condition monitoring is really justified.

5.1 TEROTECHNOLOGY

Terotechnology came into being as a word in the 1970s. The British Standard BS 3811:1993 (Glossary of Terms Used in Terotechnology) defines it as

> A combination of managerial, financial, engineering, building and other practices applied to physical assets in pursuit of economic life cycle costs. NOTE: Terotechnology is concerned with the specification and design for reliability and maintainability of physical assets such as plant machinery, equipment, building and structures. The application of terotechnology also takes into account the processes of installation, commissioning, operation, maintenance, modification and replacement. Decisions are influenced by feedback of information on design, performance and costs, throughout the life cycle of a project.

The life cycle costs are defined as

the total cost of ownership of an item taking into account all the costs of acquisition, personnel training, operating, maintenance, modification and disposal.

This is long-term planning, and a keen mind is required to appreciate it. It also requires a person who is devoted to the 'company' rather than his or her own advancement. It is not uncommon to find that in shift work the operator currently in control will conveniently fail to notice a condition warning signal so that he can achieve maximum output for his shift. This 'I'm all right, Jack' attitude unfortunately can permeate into higher levels of management with the consequent ultimate disaster (or at least loss of profits) for the company.

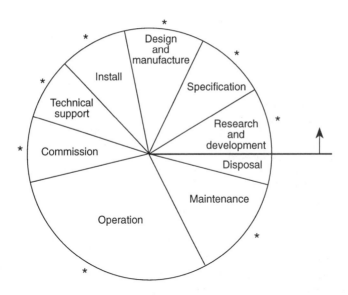

Figure 5.1 One example of the content and proportions for total life cycle costs (*, aspects where condition monitoring should be included).

Fig. 5.1 gives an idea of total life cycle costs. It does not convey what the actual costs are but it indicates where monitoring has an influence and must be considered in the cost. Starting with the research and development up to the manufacture some costs will involve 'condition monitoring' in the sense of design and fitting. From installation to commissioning the monitoring will be part of each aspect as regards the testing. The operation then includes monitoring, as will the maintenance.

The optimization of the maintenance cost is important – there can be too much maintenance as well as too little, as shown in Fig. 5.2. (This was first discussed in section 1.3.)

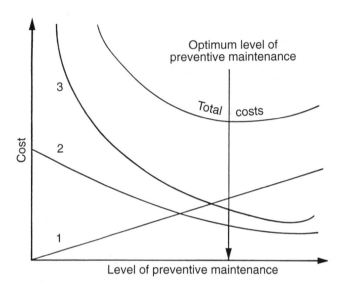

Figure 5.2 Maintenance cost depending on amount of maintenance undertaken: curve 1, preventive maintenance costs; curve 2, corrective maintenance costs; curve 3, indirect maintenance costs; the total costs curve is the sum of curves 1–3. (*Maintenance Aspects of Terotechnology*, DTI, 1975.)

What needs to be laid out then is the actual expected costs with and without monitoring.

This is not easy particularly if no previous experience of a similar machine or system is available. Table 5.1, the check list at the end of this chapter, provides a means of sorting out the various costs which may be involved. The extra build costs are those necessary because the monitoring is to be incorporated within the overall design of the machine or system. (It would be simpler and cheaper, at this stage, if there were no monitoring!) Purchase costs are self-explanatory. The running costs must include not only the consumables necessary for the particular monitor but also the human resources costs.

So far our comments on Table 5.1 have only shown the negative costs of monitoring. It is the fourth block, however, which provides the positive. Here is the *raison d'être* of monitoring – what costs will be saved?

There is a need again not only to consider hardware, but to take into account lost sales because of reduced production, extra person-hours expended, costs of replacements and even redundancy payments. The Copper Development Association (1995) lays out a series of 'costs' to an organization where electrical power problems are met, as summarized below:

- data loss costs, may be vital and irreplaceable;
- efficiency lost costs, poor process control;
- equipment costs, replacement of damaged equipment, extra protection purchased;
- downtime costs, up to thousands of pounds sterling per hour for larger manufacturing plants;
- wastage costs, production material wasted owing to uncontrolled shutdown;
- consequential losses, sales lost in market, customer dissatisfaction;
- executive time costs, time wasted in addressing problems and rescheduling work;
- safety threat, dangers to human life.

Block 5 in Table 5.1 is where monitoring is to be fitted as an afterthought. This invariably is more expensive than if fitted from the start, but it is quite often what has to be done. It must be included as a negative in the cost equation if it is the only way.

5.2 WHERE A MONITOR MAKES SENSE

A monitor, or several monitors, makes sense if the sum of the losses is less than the sum of the gains in Table 5.1. In other words, the loss to the company could well be greater if condition monitoring is not applied to the machinery or system.

Typical situations where monitoring is of great value would be the following:

- complex and expensive machinery (high cost of hardware replacement);
- machinery in remote locations (high cost of maintenance);
- personnel-dependent machinery (possible loss of life);
- continuous production facilities (high cost of restarting);
- high output facilities (high loss of sales).

One example is that of the continuous float glass process where, should one part of the 100 m run go wrong, then the whole process could take many hours to set up again owing to the complex temperature gradient which has to be arranged along the line.

An example of remote monitoring is that of mining. Large machinery, such as roof support hydraulic pumps, have to be assembled in very confined spaces and its maintenance not only is very difficult to undertake manually, the provision of spares could be kilometres away.

While the above are the more important situations for monitoring, there are many others where monitoring could be just as advantageous because it reduces overall costs. The following are a few suggestions:

- automobile   fewer breakdowns on long journeys away from base
- machine shop   less likelihood of failure during complex operations
- mobile machinery fewer failures during dangerous movements.

Long-distance coaches are one example of considerable inconvenience to passengers and crew. In Argentina and Paraguay, for instance, it is very common for tyre pressures to be remotely monitored by the driver to ensure that slow punctures do not cause complete breakdowns (or crashes) by developing too far. Pressure lines are connected to each wheel through slip-rings, and pressure gauges, one for the combined tyres of each axle, are visible to the driver (Fig. 5.3). If a drop in pressure is detected, the vehicle air system automatically pumps more air into the appropriate tyre; should a single tyre actually burst, the other tyres on the axle are isolated and the driver can stop safely and change the wheel.

Figure 5.3 The Vigia remote monitoring of coach tyre pressures, (Col-Ven S.A., Santafe, Argentina).

Another recent example is that used by Porsche to monitor the environmental condition, again to prevent crashes and more serious consequences. In this case the road surface is monitored by means of two sensor groups: one measures the amount of water on the road (structure-borne sound generated as the result of water hitting the front wheel arch) and the other is a road-texture sensor (again structure-borne sound but this time on the rear axle hub-carriers). The output from these two sensors compared with vehicle speed enables the on-board computer chip to determine the friction force-to-slip quotient.

5.3 WHERE A MONITOR DOES NOT MAKE SENSE

Perhaps the word 'monitor' is too simple here. We must remember that monitors vary between something costing a few coppers to many thousands of pounds sterling. For a monitor to 'make sense' its value has to match the cost of the machinery it is protecting. For a monitor not to make sense, it has to have a different reason. Consider the following.

- The monitor would seriously affect the operation.
- The machine is purposely meant to go to destruction.
- There is no way the output from the monitor can be assessed.
- The cost of a replacement component (and its fitting) is minuscule.

Medical machinery needs to be small in size and have a high contamination cleanliness (as regards health contamination). This may not generally allow conventional sensors to be fitted on line, but may allow off-line tests.

Test vehicles for missile practice hardly warrant monitoring for condition – it is only too obvious when they blow up.

Machinery which is slow running and of low load will last for years – and has done – without any need of monitoring and with minimum maintenance. As a confirmation, a portable monitor could be used, with a monthly reading, or even once a year.

Another condition where it is pointless to monitor is where the operation is haphazard and the maintenance slipshod. If there is no atmosphere among the staff for reliability and life and consistency and cost saving (section 1.4) then do not bother; it will be a waste of time. However, if there is a desire for the best, read on.

5.4 CHECK LIST

The check list (Table 5.1) is designed to identify what costs are involved in monitoring, i.e. the credit and the debit. It is essential to determine these costs in order to convince management of the cost effectiveness (or otherwise) of fitting machine condition monitoring. Naturally, the key result of the check is the comparison of the loss with the gain. If there is more to gain from monitoring, then it is to be encouraged.

Table 5.1 Checking for overall costs with/without monitoring

Have I checked?	*Detailed items*	*Expected costs*	
		Loss (£)	*Gain(£)*
1. Extra build costs due to the fitting of monitors	1.		#########
	2.		#########
	3.		#########
	4.		#########
	5.		#########
2. Purchase cost of monitors	1.		#########
	2.		#########
	3.		#########
	4.		#########
	5.		
3. Running costs of monitors	1.		#########
	2.		#########
	3.		#########
	4.		#########
	5.		#########
4. Costs of failures if left undetected	1.	#########	
	2.	#########	
	3.	#########	
	4.	#########	
	5.	#########	
5. Extra costs for fitting a monitor after the build of the machinery–plant (optional)	1.		#########
	2.		#########
	3.		#########
	4.		#########
	5.		#########
Is monitoring cost effective?	Total	–	+

The choice of monitor

CHAPTER 6

Input monitoring

Part Two is concerned with the choice of monitor. We have looked at how monitoring should be approached in Part One; in Part Three we shall look at the use of monitoring in practice, but Part Two is the heart of the matter – the choice of the monitor.

The 'monitor' consists of three aspects:

1. the sensing device;
2. the recording of the data;
3. the analysis of the data.

Chapter 9 will deal with the second aspect, and Chapter 10 the analysis; the first feature, the actual hardware, needs much more detail and three chapters, 6, 7 and 8 will be devoted to that subject. In order to emphasize the whole concept of monitoring, the sensing devices have been split into three types of monitoring, each associated with their positional relationship with the machine process: Chapter 6 deals with the monitoring of what goes into a machine, Chapter 7 deals with what is going on within the machine and Chapter 8 deals with the monitoring of what comes out of a machine. This chapter, Chapter 6, examines the environmental effects, power monitoring and some special input examples.

6.1 ENVIRONMENTAL MONITORS

6.1.1 Introduction

Environmental monitoring is primarily with a concern for human welfare – how quickly we will wither or wilt when our capacities are stretched. It is also now obligatory (the Environmental Protection Act 1990). However, it is also of great importance to many machine operations; in addition it can be a proactive sensor for faults which might develop. For instance it is common for most control instrumentation connected to machinery to have a range of ambient extremes between which it can operate and a slightly wider range within which it can be stored. If these limits are crossed, then the likelihood of machine failure in the future significantly increases in probability. On the monitoring of the machine operation, aspects such as visibility and temperature and any excessive escape of process gas or contaminant would be strong contenders for important on-machine monitoring (more fully covered in the next chapter, particularly in sections 7.4, 7.5 and 7.9). Efficiency of processes, where the combustion gases are monitored, is part of the discussion in Chapter 8.

6.1.2 Brief description

The type of environmental condition which can affect machinery, and which can be monitored, would be such as (Fig. 6.1)

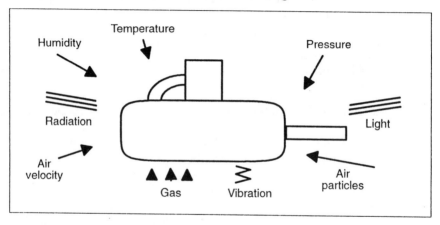

Figure 6.1 Outside influences on a machine.

- ambient temperature,
- relative humidity,
- barometric pressure,
- light intensity,
- air velocity,
- airborne particulate,
- radiation,
- escaped gases.

6.1.3 Practical applications

Typical effects of the environment on machine condition are indicated in Table 6.1.

Table 6.1 Some effects of the atmosphere on machinery

Atmospheric condition	Effect on machinery
Ambient temperature	Both extremes are important. Too low a temperature can seize components or make them inoperable, or freeze the fuel. Too high a temperature can cause distortion of components, break-up of insulation, evaporation, burning etc.
Relative humidity	If this is excessive, say, over 80%, corrosion will be a problem. There is a possibility of dilution of lubricants with condensation.
Barometric pressure	Some sensors may need checking if the barometric pressure is too low or too high.
Light intensity	The importance of light intensity is two-fold. It relates to first of all the visibility of the machine operations (and displays) to an operator, and secondly, the permanent effect that the light rays have on machine 'fabric'. VDU displays can sometimes be misread if the light intensity is too high – and this can cause machine failure. The permanent light effects would cover such aspects as fading and perishing.
Ground vibration	Vibration is not strictly an environmental feature, but certainly vibration can be transmitted through the environment to cause a problem at a machine. This is not only the earthquake tremor; it is also the shaking induced from an out-of-balance machine which may be in the vicinity and on the same base foundation.
Air velocity	Perhaps a little obscure, but gusts of wind or draughts can cause failure in reading instructions.
Airborne particulate	Contaminant ingestion into machines is a major problem with lubricated and hydraulically operated machinery; this involves not only solid particulates but also airborne bacteria and fungi.
Radiation	Radiation which is electromagnetic can cause total confusion to some electronic control unless properly shielded. This is known as EMC (electromagnetic compatibility), EMI (electromagnetic interference) or RFI (radio frequency interference).
Escaped gases	The main gas problems relate to corrosion or reduction in liquid density.

6.1.4 Detailed discussion of environmental measurement

Each sensor is normally an individual sensor just monitoring one of the conditions mentioned in section 6.1.2; however, a number of them can be combined by being connected to one control unit (Fig. 6.2 shows an example of this).

Within the control unit would be an additional means of storage and a means of output to another analyser such as a PC. Alternatively, or in addition, the analysis and display of the instant results or trends may be

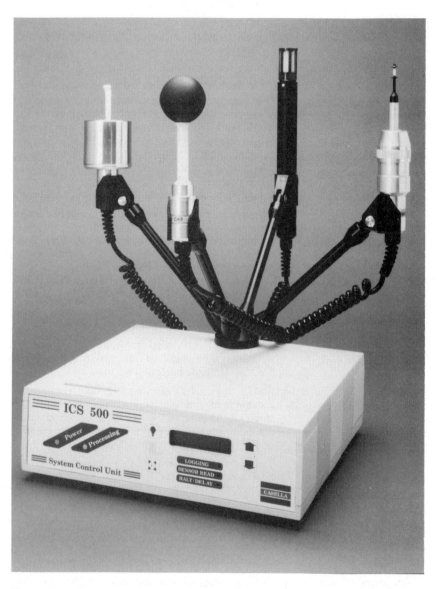

Figure 6.2 Indoor environmental sensing (Indoor Climate System ICS500, Casella).

included. The automatic recording at regular intervals (such as every few seconds, or every hour, or once a day) is a valuable feature.

The monitors are possibly best described as 'sensors'. The following list describes some of the features of each and the types of sensors and their scope.

(a) Ambient temperature

Range of temperature monitored would normally be from, say, −30°C to 80°C although a much increased top end is possible for instruments in a non-human environment (perhaps 185°C).

There is a wide range of temperature sensors available; most of these are for the machine processes rather than the input to the machine, but for completeness the majority of types are shown on one table, i.e. Table 7.26. Which is used will depend on the control available and the application.

(b) Relative humidity

Relative humidity is the ratio of the water content in the air to the water content necessary to cause condensation (saturation). It is stated as a percentage; it is also dependent on the temperature of the air (Fig. A.4).

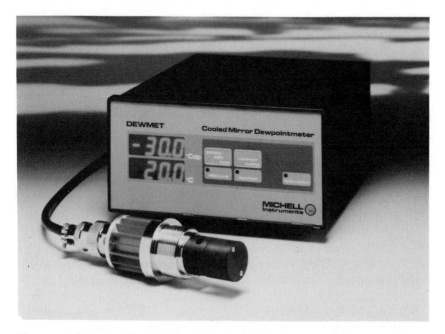

Figure 6.3 Dewmet dewpoint transmitter and meter (Michell Instruments).

'Dewpoint' looks at the subject in the other direction; it is the temperature at which the water vapour present in the air saturates the air and begins to condense (i.e. dew begins to form). An example of a dewpoint instrument is shown in Fig. 6.3.

Traditionally relative humidity has been determined by the 'wet and dry hygrometer' where two thermometers, side by side, with one constantly moistened, show a difference because of evaporation. A later technique is that of using a 'chilled mirror' to obtain the optical dew point.

Today the measurement of relative humidity is undertaken on line with one or another means of sensing moisture. This should be able to be carried out in adverse environments which might include process contaminants, ash and soot.

(c) Barometric pressure

Barometric pressure is measured as a standard by a column of mercury, but the aneroid (without liquid) barometer is more familiar these days (it is considerably smaller, but it does need calibrating and checking from time to time). A pressure of 1 bar is equivalent to 760 mm of mercury (29.9 in).

(d) Light intensity

Luminous intensity (illumination) is measured in 'lux', which is the illumination of one 'lumen' per square metre. Lux meters can measure from below 1 lux to many thousands, e.g. 50 000. While the eye is normally able to cope with such a wide range, not all machinery displays remain visible throughout.

(e) Ground vibration

Vibration sensing and analysis are dealt with in great detail in the next chapter. As regards earthquake monitoring a seismic device is necessary; this is not normally expected in the UK but could be essential in other lands.

(f) Air velocity

The velocity of air is measured in simple terms by a wind sock, i.e. the angle at which it hangs. More precise measurement is made with a variety of instrumentation, for instance the following examples:

- $> 2.5\,\mathrm{m\,s^{-1}}$ – Pitot static tube;
- $> 0.25\,\mathrm{m\,s^{-1}}$ – rotating vane–propeller (average over vane area);
- $> 0.10\,\mathrm{m\,s^{-1}}$ – thermal anemometry (for smaller regions).

Air flows of the order of up to $50 \, \text{m s}^{-1}$, with an accuracy of 5% of read-ing, can be determined by means of such portable instrumentation. (In passing it should be mentioned that where the air contains particulate, or is a mixture of gases such as in a flue gas, ultrasonics can be used, per-haps up to velocities of $85 \, \text{m s}^{-1}$.)

(g) Airborne particulate

The main machine problem with particulate is from solid particles. As such they are detected by airborne particle counters covering perhaps a particle size range from $0.3 \, \mu\text{m}$ to, say, $25 \, \mu\text{m}$ with con-centrations up to around $2.5 \times 10^7 \text{m}^{-3}$. Two of the better known types are optical (laser) particle counters and aerodynamic particle sizers.

For the more concentrated situations gravimetric or dust monitors are used. The disadvantage of the otherwise very accurate gravimetric types is that only an average dust concentration is determined over a period of time (as measured on a collector membrane). Portable dust monitors can use a laser light scattering technique (at, say, 70° or forward narrow angle) and measure up to $200 \, \text{mg m}^{-3}$ and provide a collection of the particles after the instantaneous measurement. It must be understood that the two are different in that one measures weight, the other shape–size; a comparison can be made if the density of the particles is known.

(h) Radiation

Radiation may be electrically induced or may originate from some other source, such as from radioactive isotopes. However, the term usually applies to the emission of electromagnetic waves. These waves occur at the speed of light ($v \approx 3 \times 10^8 \text{m s}^{-1}$) with a range of wavelengths cover-ing from γ rays to radio waves, (note that $f = v/\lambda$) (see Table 6.2). It should be noted that electromagnetic waves do not need a medium through which to pass (unlike sound waves). The higher the frequency, the greater the penetration.

Control instruments (and measurement instruments) need to be able to function correctly in the presence of electromagnetic waves, i.e. to be compatible with the situation. This may mean shielding the sus-ceptible components. EMC emission is normally only concerned with frequencies above $10 \, \text{kHz}$, and could reach a maximum of $1000 \, \text{MHz}$ (i.e. 10^9Hz). There are a number of standards and methods associated with the measurement of EMC, and books about electricity should be consulted.

Table 6.2 The electromagnetic spectrum

Type	Wavelength $\lambda\ (m)$	Frequency $f\ (Hz)$	Production	Comments
Radio	$>10^{-4}$	$<3 \times 10^{12}$	Electrons oscillating in wires	This can affect electronic equipment
Infra-red	$7 \times 10^{-7} - 10^{-4}$	$3 \times 10^{12} - 4 \times 10^{14}$	Hot objects	Better as a sensor
Visible	$4 \times 10^{-7} - 7 \times 10^{-7}$	$4 \times 10^{14} - 7.5 \times 10^{14}$	Very hot objects	Visual colour range
Ultra-violet	$10^{-9} - 4 \times 10^{-7}$	$7.5 \times 10^{14} - 3 \times 10^{17}$	Arcs and gas discharges	Senses fluorescence
X-rays	$10^{-12} - 10^{-8}$	$3 \times 10^{16} - 3 \times 10^{20}$	Electrons hitting metal targets	Detection of internal flaws
γ rays	$<10^{-10}$	$>3 \times 10^{18}$	Radioactive nuclei	Pre-treatment of wearing substance for detection by Geiger counter

(i) Escaped gases

Historically important for the impact on human beings, and still used in 1995, is the 'canary in the cage' detector. However, although the canary may be the absolute for lethal gases, there are now many detector tubes available for specific gases. There are also scent detectors, or sniffers, which look for the unusual in atmospheric gaseous content.

Infra-red radiation may be used; in this case an IR beam passes through the gas and either the amount absorbed (in comparison with an acceptable reference gas) or the different wavelengths affected indicate the presence of specific gases, i.e. the beam wavelength is tuned to the absorbing frequency of the gas. A similar arrangement is available using an ultra-violet beam. IR is used for, say, CO, CO_2, CH_4 (methane). UV is used for, say, SO_2, H_2S and O_3 (ozone).

As well as the IR and UV techniques (including Fourier transform infra-red and non-dispersive infra-red), there are many other analytical techniques. Further details are given in section 7.1.4(b) on smell monitoring with a comprehensive list of techniques in Table 7.3.

6.1.5 How to start

Section 6.1 has introduced the various possibilities. This should enable the user to have a good idea of what might be necessary before contacting a supplier. If the supplier does not cover the particular type in mind then try another supplier.

Temperature monitoring is well advanced, and most suppliers should be able to supply the required type, and the expertise needed, without recourse to a pre-trial.

6.2 POWER MONITORS

6.2.1 Introduction

The monitoring of power exerted or expended – in comparison with the output – is the key aspect of efficiency monitoring (discussed in section 8.1). However, it also has a significant contribution to condition monitoring in its own right; for instance, a wide variation above the normal, or below it, will signal a fault in the system – additional power going to friction or temperature, reduction in power with free running (due to a coupling failure) etc.

The immediate power source may be pneumatic, hydraulic or mechanical, or further back in the cycle, the electricity or fuel consumed, or head lost. The monitoring would thus be directly related to one of these (Fig. 6.4).

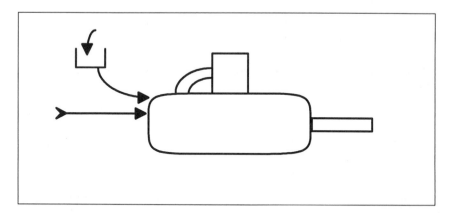

Figure 6.4 Power monitoring

6.2.2 Brief description

Power is monitored to determine the condition of the machine and to sense when more or less power is needed owing to a fault.

6.2.3 Practical applications

Any situation where power is being expended is suitable for this monitoring. It is particularly relevant to electrical motors which are providing the basic power source, because this can provide an instantaneous signal when a fault develops; other power consumptions may take longer to assess (e.g. kilometres per litre on a motor vehicle). While complex analysis can indicate a range of faults (discussed below), a simple change in electrical current provides an immediate indication of a load change; this is applicable to many applications, e.g. the breakage or wearing of a cutting tool.

6.2.4 Detailed discussion of power measurements

The assessment of fuel expended per exercise or operation is a fairly simple calculation. The monitoring of electrical motors, however, is much more rewarding and can provide the data required for system monitoring in very short times with an indication of faults at the time they are occurring.

By analysing the frequency spectrum of one phase of the current input to the motor a considerable amount of information can be gained on the motor condition, both electrical and mechanical. In other words, by clamping a current transformer, or a Hall effect probe, on the appropriate lead (or using a current supply tapping like the motor ammeter circuit) the current signal is analysed by a spectrum analyser (much as described in section 7.2, on vibration analysis). The slip frequency and the rotational frequency, and the side-bands and harmonics associated with them, provide sufficient information to diagnose such motor problems as

- mechanical out-of-balance due to bearing faults,
- wear of stator bore or rotor,
- bent shaft due to thermal problems,
- broken rotor bars and
- cracked rotor end rings

as well as the more electrically associated problems with windings and joints. (Using search coils on the motor ends will enable the change in any axial flux induced in the coils to be monitored.) Figure 6.5 is one example of a power monitor which covers a range of features.

Electrical power monitoring has an additional value in that the prime mover itself may be at fault. It is estimated by Fluke International that between 15 000 and 30 000 sites in the UK are currently exposed to serious power quality problems. Developing this thought further the Copper Development Association consider that around £200 million of computer-related 'consequential loss' each year is caused by poor power quality. Harmonics, earth leakage and voltage disturbances are involved.

Figure 6.5 Motor power analyser (Fenner 360).

Motor load can be monitored from the current drawn, as shown above (see also section 7.7.3). However, for analysis of the system an earlier indication of load deviation can be obtained from a monitoring of the shaft torque either directly or by calculation of electrical parameters. The torque can be detected by non-contact FM or inductive transmission of data based on a strain gauge pattern; other means could be the examination of the surface distortion optically, or an in-line torque meter. The change in torque should be able to detect such features on a pump as

- blockages,
- dry running,
- cavitation,
- change in fluid density and
- high bearing temperatures.

Unlike the detection of high current monitoring, torque monitoring is able to cope with high gear ratios and very slow speeds with a sensitivity as low as 1% of full load. The torque can be calculated from the applied three-phase voltage and the line current, and compared with the full load torque.

6.2.5 How to start

Fuel consumption should be a standard recorded figure; it can be used if the operation is fairly stable and repeatable.

Where electric motors are used, check to see whether a torque device can be fitted on the motor shaft (it is easier than most people believe); if not, consider a current measuring device or watt meter.

6.3 MONITORS IN SPECIAL CIRCUMSTANCES

6.3.1 Introduction

This brief section is to encourage further lateral thought. It is relatively easy to decide to fit one of the well-known monitoring devices, but there could be a far better solution from a combination of simple ideas. The whole concept of CME (Chapter 2) is to develop a thought process which arrives at the most cost-effective reliable system by means of the 'best' monitor.

The 'C' (components which fail) is highly important. Do I have a special system where failures are occurring (or might occur) which are out of the ordinary? If so, maybe I have special circumstances which warrant a special monitor.

6.3.2 Brief description

A monitor or monitoring system is developed which is specific to a special circumstance or operation (Fig. 6.6).

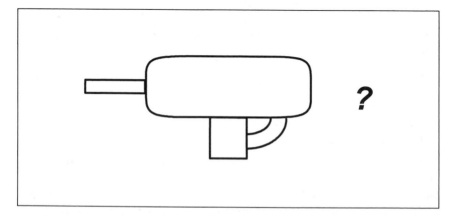

Figure 6.6 What other input detects the machine condition?

6.3.3 Practical applications

Special circumstances are usually in a 'one-off' situation; the individuality being either the machine or where it is used. The industry is irrelevant, it could be aerospace or the zoological garden, a multimillion pound consortium or a small workshop. The important thing is that it is something occurring in the machine world which is different.

6.3.4 Detailed discussion

As an example consider the following case from underground coal mining

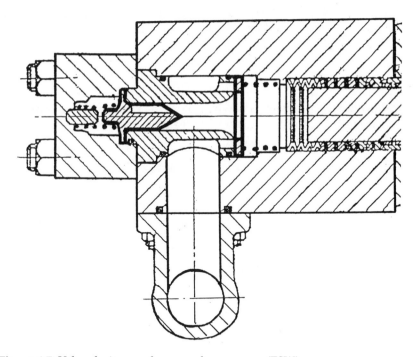

Figure 6.7 Valve design on slow speed ram pump (FSW).

A slow-speed ram pump (Fig. 6.7) had been failing owing to the break-up of its valves. A means of monitoring was required. A number of conventional monitors would come to mind such as

- vibration, to sense the change in excitation when the valve became unstable,
- wear debris analysis, to sense when particles began to be produced from the valve surface,
- temperature, to sense when the casing began to rise owing to excessive friction, or
- thermodynamic efficiency, to sense when the output began to drop.

However, by examining the valve dynamics it was apparent that because the pump was being used underground the inlet pressure could very easily drop below the permissible level (the fluid reservoir was in the same mine shaft, at the same level, as the pump). Fluid level could drop or the outlet from the reservoir become partially blocked and the inlet pressure drop below atmospheric with complete instability of the valve and eventual break-up. The solution, therefore, was to place an inlet monitor in the line, for the simple purpose of just monitoring the inlet pressure. This basic expediency was then able to prevent running of the pump until the fluid level had been topped up or any obstructions removed.

6.3.5 How to start

Think around the problem. Just why is it there? What are the parameters which are influencing the situation? Have a brainstorming session.

CHAPTER 7

On-machine monitoring

The monitoring of the machine itself – the on-machine monitoring – can be considered in two ways:

- direct human monitoring – human;
- direct sensor monitoring – automatic.

While 'human' is entirely the responsibility of a human being using his or her own senses, the 'automatic' may or may not involve human intervention. For instance it is quite feasible (and very common) for a sensor output to be fed directly to a computer where the decisions are made. On the other hand, where the circuit is less complex, it can be quite in order for the human to examine the sensor output and make the decision.

In the following nine sections specific types of both manual and automatic monitoring will be discussed. They are totally different and tend to detect quite different machine conditions. In some cases there is an overlap in detection but in most cases the techniques are completely independent:

1. human;
2. vibration;
3. ultrasonics;
4. wear debris analysis;
5. oil analysis;
6. thermography;
7. leakage;
8. corrosion;
9. steady state analysis.

There is a considerable range of manufacturers and suppliers for all these types of monitors – a list of some of them is given at the end of the book to provide the user with a selection with which to commence. However, the names given are not necessarily the best although their representatives may have been kind enough to co-operate with the author. The author would also be pleased to add others to his database if manufacturers or suppliers would be kind enough to send details.

7.1 HUMAN

7.1.1 Introduction

By 'human' is meant all human observation, where one or more of the five human senses have been used. (The monitoring of the human has been discussed in section 1.5.) While such human observation of the condition of a machine is most likely the result of many years' experience by a skilled worker, there are certain aspects which a novice can quickly appreciate. Having achieved a certain understanding of what monitoring is all about – from Part One of this book – most engineers with limited experience, but using their physical senses, should be able to notice certain changes occurring in machinery.

While this section is primarily to do with unaided human assessment, two features which can be assisted by instrument are also mentioned – colour and smell. A few other aspects which do have very close ties with a range of instruments are discussed later in this chapter.

7.1.2 Brief description

Human monitoring is the direct use of the human senses to sense a feature which reveals a fault developing in a machine (Fig. 7.1).

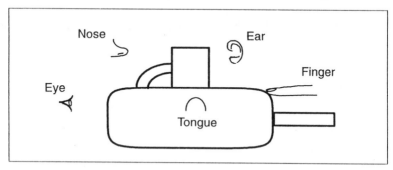

Figure 7.1 Human sensing of machines.

7.1.3 Practical applications

This is simple low cost monitoring – initially. However, a person employed to look or listen can become very expensive after a while, so such monitoring should be replaced after it becomes clear that condition monitoring is here to stay. It is, however, the first step for most machinery applications.

7.1.4 Detailed discussion of measurement by humans

Table 7.1 gives some practical examples where the human senses have been very valuable in monitoring machinery.

Human monitoring has its place in the early stages of appreciation of machine monitoring. It is, however, limited because of being so subjective and dependent on the condition and experience of the human. It is also limited by a very time-consuming, operator-intensive, means of logging the data for later analysis. Because detection devices have developed so well in recent years there is no longer a need for this limited method although, occasionally, a highly skilled person may just come across the odd failure mode developing which has been missed by the fitted automatic detectors.

There are two types of monitors, which directly relate very closely to the human monitoring, which are not detailed elsewhere in this book. They are, however, appropriate for machine monitoring in certain circumstances so they are briefly discussed here:

- colour monitoring;
- smell monitoring

(a) Colour monitoring

The human eye is designed to be able to differentiate between up to 10 million colour hues in good lighting. Automatic monitoring is not as good but has the advantage of not varying from day to day; in other words, it is more consistent in quantifying colour. A reflectometer will detect the luminous reflectance or gloss of surfaces.

Colour in liquids is normally made up of either a soluble dyestuff or an insoluble pigment. In order to determine these with any accuracy it is important to remove other solids in the liquid which might mask the real 'colour'; a filter (possibly $0.45\,\mu m$ membrane) is suitable.

Colour of water is an issue all on its own. The intensity of such colour is measured in Hazen units where Hazen is almost equivalent to turbidity at a chosen light frequency (e.g. 390 nm); however, the true turbidity associated with the solid particulate present has to be subtracted (e.g. at 650 nm). Water colour is primarily cosmetic – does the customer like the look of it? In machinery, however, the colour of the liquids can convey what has happened to a fluid because of the machinery malfunction (as mentioned in Table 7.1).

Table 7.1 Human monitoring

Sense	Noticeable change	Some possible causes
Eye	General appearance (e.g. leakage)	Fractured pipe Displaced joint Defective seal Loose connector
	Macro movement (e.g. juddering)	Out-of-balance due to part failing Assembly or housing loose Defective damper Support fractured
	Surface colour	Excessive temperature due to lack of coolant or no lubricant between sliding surfaces
	Liquid colour	Oxidation due to running at too high a temperature.
	Liquid opacity	Excessive wear Failed air filter
Ear	External mechanical sounds	Loose bolt, mounting or pipe Loose or defective coupling Broken shaft
	Internal mechanical sounds	Bearing damaged Faulty gearing Insufficient lubrication Wrong speed
	Flow sounds	Cavitation Leakage High pressure seals defective Excessive air due to low fluid level
Nose	Smell	Liquid oxidation High temperature running
Finger or body feel	Temperature	Inefficient operation due to wearing surface Defective solenoid Pipe constrictions
	Vibration	Fatigue of gear Out-of-balance or misalignment Defective valves Air inclusion in system Loose parts
	Moisture	Leakage from seals, pipes or reservoir Faulty gauges Covers loose
Tongue	Not recommended owing to health hazards	–

The three primary monochromatic colours (RGB) which make up white light are

- red, 700 nm,
- green, 546.1 nm, and
- blue, 435.8 nm.

Because these colours are mutually exclusive, the amount of light which passes through a set of red, green and blue filters will identify the colour. This is the simplest way of monitoring colour and, although it is usually preferred for colour matching rather than for precise detection, it can still be effective for machinery. The other major way is to pass the light through a diffraction grating or prism and to use a spectrophotometer to analyse the resulting spectrum and to produce an acceptable 'signature'. Once a signature has been obtained (either with three colours or the spectrum) is is relatively easy to undertake a condition monitoring exercise relating each new spectrum obtained to the original signature. The complexity of the signature need not be very great, for instance four or even just the three primary frequency points may be quite adequate to show up the differences in machine condition.

Table 7.2 Colour monitoring

Type of monitor	Technique	Comments
Visual comparison	The use of carefully prepared colour charts which are matched by eye	This can only be done when illuminated by the stated light source Can vary slightly from person to person
Absorbance (at a wavelength) or transmittance	Each colour has a specific wavelength which may be absorbed (Fibre-optic based sensors and colorimeters)	The usual visual wavelength range is from 400 nm to 650 nm with 550 nm the most sensitive to the human eye Probably more useful as a colour matching technique
Scanning spectrophotometry	Spectrum analysis via a diffraction grating or prism	Accurate but expensive Less expensive is where an array of photodiodes is used
Tristimulus co-ordinates	The colour is described in terms of eight co-ordinates – see also Fig. 7.2	Helpful for accurate comparison of colours

Table 7.2 outlines different techniques. Detection of the reflected ambient light and detection of responses from specific light sources are both used. Because reflectance may be influenced by the ambient light, care should be taken to ensure the device works in the situation being monitored, e.g. from 1 to 1000 lux.

The 3-bit coding of colours, based directly on the three primary colours, is given as follows:

- 000, black;
- 001, blue;
- 010, green;
- 011, cyan;
- 100, red;
- 101, magenta;
- 110, yellow;
- 111, white.

Two examples of in-line colour monitors are shown in Fig. 7.2 and 7.3.

(b) Smell monitoring

The automatic monitoring of smell has been notoriously difficult over the years – partly because of the competition from the human nose and even more from certain animals which totally outstripped chemical 'sniffers' (section 6.1.4i). However, that is now beginning to change gradually with various devices able to differentiate quite well between one (acceptable) smell and another (unacceptable because it is just off) smell (levels of parts per billion are now achievable). This will be particularly valuable in process applications but as the price drops may well be suitable for other oil or liquid applications.

Electrochemical sensors can be obtained for specific process gases, and each such industry will be aware of what it needs for its application. In the case of ozone, for instance, one sensor is an electrochemical cell in which a gold cathode, silver anode and aqueous potassium bromide electrolyte are isolated from the sample to be tested by a gas permeable, water impermeable, membrane. A list of different devices is given in Table 7.3.

Gases, in general, started being analysed in the early 1920s. Gas detection tubes, such as Draegertubes, are glass tubes slightly smaller than a pen, filled with a chemical which changes colour in the presence of a given gas or vapour; over 150 types of tube are available to cover 350 different types of gas and vapour. These devices can indicate the concentrations of the gas. The Gastec tubes can be simply used to draw a

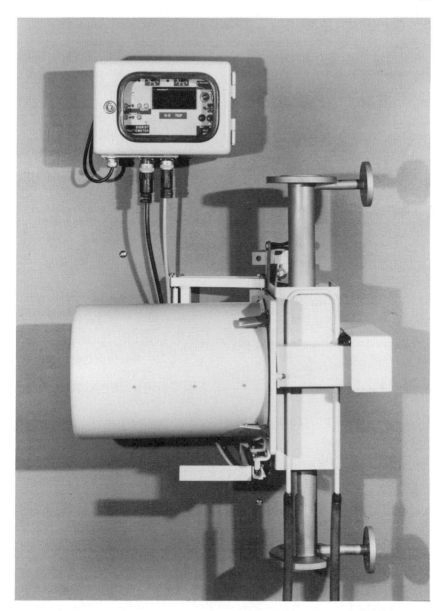

Figure 7.2 Full in-line colour monitor (Sigrist).

quantity of gas in and immediately read the concentration for the specific gas tube chosen. Somewhat different, but providing the same service, are Chemcassettes; in this case the chemically prepared pads change colour with the intensity of the gas they are to detect; the colour can then be checked by spectrophotometry.

Table 7.3 Smell and gas monitoring

Type of monitor	Technique	Comments
Electrochemical	Electrochemical reaction	Each arrangement is specific to one type of gas
Chemical colour	Chemical changes colour in presence of gas	Each tube or pad is specific to one type of gas
Metal-oxide semiconductor	Electron density modification	With suitable software (and possible combination with polymers) a range of smells can be detected and identified
Conducting polymers	Polymer interaction with volatile chemicals causes a change in electrical resistance	An array of polymer sensors can detect a range of different gases; 12 at a time with up to 25 available appropriate to industry
Biosensor	Inert biological materials which produce a millivolt output	Suitable for organic and work-shop smells; highly specific
Mass spectrometry	Ionization of the gas examined spectrometrically	A database can be set up with known patterns of gas compounds
Solid electrolyte	ZrO_2	Suitable for detection of O_2 in flue gases
Fourier transform IR	Scanning by IR	Not suitable for single element gases; wide range of gases (perhaps 12)
Non-dispersive infra-red	IR absorption at specific wavelengths (i.e. single gas filter)	Not suitable for single element gases; used for CO, CO_2, NH_3, CH_4, SO_2, etc.
Non-dispersive ultra-violet	Similar to IR absorption	More specific than the IR absorption; used for SO_2, NO_2, H_2S, Cl_2 and O_3.
Paramagnetic	Detects molecules which exhibit paramagnetic properties	Suitable for O_2
Flame ionization	Special detector (similar to gas chromatography	Not suitable for single element gases; hydrocarbons and VOCs
Chemiluminescence	Light emitted from chemical reaction between sample and ozone, measured by photomultiplier	Highly specific to nitrogen oxides (NO_x)

Figure 7.3 Small sensing head RGB colour device (OEM).

Figure 7.4 The AromaScanner for detecting smell patterns down to parts per billion levels (AromaScan).

Sensors based on the human nose use olfactory metal-oxide-semiconductor sensors which are used to trap molecules of an odour-generating substance; an electron density modification then occurs. Each chemical compound or mixture has a specific curve (as regards shape, slope etc.) which relates to the intensity of the odour. One type simultaneously uses a hybrid of both semiconducting oxide sensors and conducting polymers and a neural network to produce an ever-improving pattern of a particular smell – an even closer approximation to the nose, with a sensitivity down to parts per billion. One example is shown in Fig. 7.4. A typical display pattern of 20 sensors showing a variation over time is shown in Fig. 7.5.

Biosensors are another means which are very suitable for organic vapours and general workshop smells. Currently the analysis of the results is expensive (with the use of mass spectrometers) but once that cost is reduced this technique could well be valuable for general machine condition monitoring.

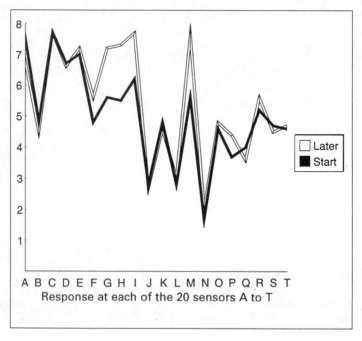

Figure 7.5 An example of change in smell pattern with time (20 sensors).

The human nose alone can be very effective when entering a region which has perhaps been running overnight with little ventilation. On one occasion because of running at excessive temperature the hydraulic fluid had exceeded its maximum operating temperature and had partly oxidized. The pungent smell was the first indication.

7.2 VIBRATION

7.2.1 Introduction

Vibration is the oldest type of machine monitoring technique. One of its earliest uses, as regards observation, was the wheel tapper who checked whether cracks had developed in a locomotive wheel by striking it with a 'sledge' hammer and listening to the sonic response. Today, not only has the recording of the sound emanating from machinery become very refined but also the analysis has progressed into a variety of highly complex techniques. The sensing and the analysis will be discussed separately in this section.

Normally the range of frequency of such oscillations decides the type of sensing and analysis. The following are the main frequency band designations with an approximation to the range of frequencies covered (Hz, hertz, is cycles per second):

- vibration, <1 Hz–25 kHz;
- ultrasonics, 20 kHz–100 kHz;
- shock pulse, 32 kHz;
- stress wave, 100 kHz–1 MHz (also acoustic emission).

In this section only the 'vibration' range is discussed. All the higher frequencies are covered in section 7.3.

7.2.2 Brief description

Vibration monitoring (Fig. 7.6) is the detection of the oscillation of a surface or structure up to 25 kHz. While most machines vibrate to a certain degree, it is the change in the vibration spectrum (range of frequencies) and the amplitude of the vibration which can indicate faults developing.

The human ear at its optimum (late teens or earlier) may just be able to sense 20 kHz, but nearer 17 kHz would be more typical. At the lower end we tend to differentiate between a 'vibration' and a 'pulse rate', the pulse rate tending to be a 'feeling' rather than a 'hearing' (below about 25 Hz).

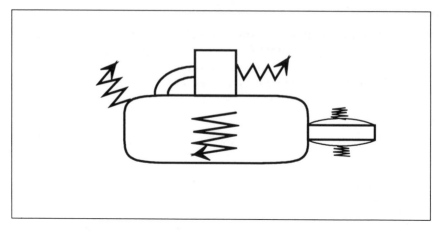

Figure 7.6 Vibration monitoring.

7.2.3 Practical applications

All machinery with moving parts can generate vibration and hence is suitable for vibration monitoring. In particular, reciprocating and rotating machinery is ideal. In addition, there are the features which oscillate because of external excitation – forced vibration – and this would include more traditionally stable items such as bridges which can be influenced by wind. Common machine features which are monitored are

- gears,
- bearings,
- shafts and
- panels.

7.2.4 Detailed discussion of vibration monitoring

(a) Vibration sensors

Vibration can be sensed by a device which is distorted or changed in some way when vibration occurs. It may sense the displacement, the velocity or the acceleration.

- Low frequency is best sensed by displacement (amplitude).
- Medium frequency is best sensed by velocity.
- High frequency is best sensed by acceleration.

Because of the reasonably large range of frequencies that need to be analysed, the low frequency displacement transducer is rarely used except for those cases where a single low frequency is being investigated. One example of this would be an investigation into the eccentricity of a rotating shaft; this is particularly appropriate because the sensor is not affected by other components close at hand.

Generally the best measurement overall is achieved by detecting acceleration. For this the simplest form would be a strain gauge stuck onto the surface which is likely to vibrate. However, a strain gauge is not too convenient to fit quickly and reliably – it requires curing and sealing – and hence accelerometers are more common. Accelerometers are usually small cylindrical capsules containing a free mass (seismic mass) bonded via a crystal to a central pillar fixed to the machine; as the little capsule is shaken by the vibration so the internal mass causes a signal to be generated by the crystal. Some gauges, such as piezo-resistive pressure sensors, are able to detect steady (0 Hz) movement as well as

Figure 7.7 Two miniature accelerometers (Entran).

the oscillation. Two miniature accelerometer types are shown in Fig. 7.7; these weigh less than 0.5 g and are less than 4 mm × 4 mm × 7 mm in size, including a 250 mV full-scale output requiring no additional amplification.

Two important features of the accelerometer need to be appreciated.

1. It has its own natural frequency.
2. It needs to be firmly attached to the item it is sensing.

Suppliers of accelerometers will state the natural frequency, often termed the 'mounted resonance frequency'. The useful working range of the accelerometer is well below this frequency, such that no greater level of error than, say, 5% will result. The higher this frequency is, the lower will be the sensitivity at the bottom end of the scale.

The output of the accelerometer can be affected by several atmospheric conditions, such as temperature, humidity, radiation, magnetism, sound levels etc., and its sensitivity to these should be checked. However, it is the way it is mounted which is likely to cause the most error. A permanent stud fitting is possible with some; others have to be stuck on – preferably to a good flat surface so that connection is made over a reasonable area. It must also be pointed out that the accelerometer must be small enough so as not to change the vibration characteristics of the surface.

Vibration can also be sensed by a microphone. In this case the transmittance of the vibration is not through a solid structure but through air (as noise) or another medium. One problem with vibration detection is the possibility of detecting vibration from more than one source at a time; with microphones this can be even more acute owing to the lower directional sensitivity and the damping in the transmitting media. For effective trend monitoring of vibration the sensing must be undertaken in a repeatable manner, both in sensitivity and in direction; sound, therefore, while being a useful human detection media, is not so effective for machine monitors. The measure often quoted of decibel is, however, important not just for sound but also for vibration.

The **decibel** (dB) is the ratio of one amplitude to another expressed in logarithmic form. For vibration, the following relationship exists:

$$N \text{ (dB)} = 10 \log_{10}(a^2/are^2) = 20 \log_{10}(a/are)$$

where N is the number of decibels, a the measured vibration level and *are* the reference level. (According to ISO 1683, the reference levels are to be taken as follows:

for acceleration, 10^{-6} m s^{-2}; for velocity, 10^{-9} m s^{-1}; for displacement, 10^{-12} m.)

g, the acceleration due to gravity, is another common way of describing the alternating forces applied to accelerometers. The gauges in Fig. 7.7 are available with full scale between 5 g and 5000 g. ($g = 9.81 \text{ m s}^{-2}$.)

Another effective means of detection is possible through optical methods. Light, particularly laser light, reflected from a vibrating surface is not usually confused by other surface vibrations. (The exception to this rule is where the laser equipment itself is being shaken by another source.) This type of detection has another distinct advantage – that of being remote and not needing attachment to the surface, although to be consistent a part of the surface needs to be defined in some way (e.g. by means of something reflective fixed permanently to the surface).

Because of the nature of the sensor outputs, the vibration can either be analysed directly, in real time, or be recorded on a tape recorder, disc or chip for later analysis.

(b) Vibration analysis

This book does not go into the theory of vibration but nevertheless we can still appreciate the subtle differences in some of the quite complex techniques. More important, though, is to sense why one technique is better, or more appropriate, than another.

Before looking at the different methods it is important to know what is meant by Fourier analysis. Fourier discovered that all complex vibration

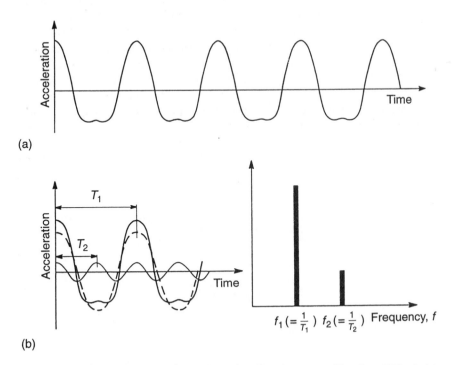

Figure 7.8 The Fourier transformation of a vibration wave (Bruel and Kjær): (a) non-harmonic periodic motion (piston in a combustion engine); (b) signal transformed into separate frequencies, in time and in frequency.

curves (level against time) could be broken down into many simple sinusoidal curves (each of one frequency with one amplitude). Hence by doing
a 'Fourier analysis' a complex wave could be broken down into a variety
of levels (amplitudes) at a variety of frequencies. In effect, the vibration
level against time has been transformed into a constantly changing ampli-
tudes-against-frequency display. The process by which this is performed
is now called a 'fast Fourier transform' (FFT) (Fig. 7.8).

Six techniques are discussed briefly followed by their practical
advantages and disadvantages:

- level of RMS general vibration against time;
- averaged level against time and number of cycles;
- level of peaks against frequency (frequency);
- level of side-bands against frequency;
- level of harmonics against frequency;
- level of harmonic bands against quefrency (cepstrum).

RMS analysis (Fig. 7.9)

The root mean square (RMS) value of a vibration signal is a measure of
the power content of the vibration. It is a simple value but over a given
length of time can be very effective in detecting a major out-of-balance
of a rotating system. It is, after all, suited primarily to a single sinusoidal
wave, rather than a complex wave, and hence out-of-balance is very
appropriate. 'Crest factor' is the ratio of the peak value to the RMS value
and for normal operation may reach between 2 and 6; above 6 will mean
that problems are developing (probably in the bearings). It is not
considered a very sensitive technique.

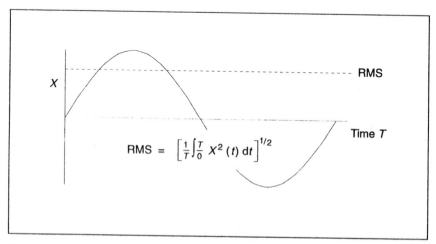

$$\text{RMS} = \left[\frac{1}{T}\int_0^T x^2(t)\, dt\right]^{1/2}$$

Figure 7.9 The RMS value.

Time-averaged analysis (Fig. 7.10)

Most real mechanical systems produce a slightly varied signal with each rotation. (Statistically this is termed 'stochastic', in comparison with the identically repeated signals which are 'deterministic'.) The more refined the fit of sliding and rolling parts, the less the variation, but nevertheless there is a variation, and with the majority of systems that difference can be so high that it masks any changes due to a fault developing. The presence of random noise can also confuse the signal. By starting an average level check at precisely the same part of the cycle of rotation, and averaging over a number of cycles, a very clear 'real-time' wave is produced. It is devoid of the random signals and it will show whether there is one part of the cycle which is changing more than another, e.g. one piston in a multipiston pump.

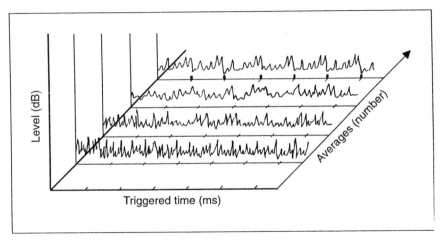

Figure 7.10 Improved signal with time averaging.

Frequency analysis (and waterfall plot, Fig. 7.11)

Every piece of machinery which has moving parts produces a spectrum of frequencies. A rotating gear wheel will produce an amplitude at the tooth meshing frequency as well as the rotational frequency. Bearings have many frequencies owing to the different diameters of the rolling elements (see later). The amplitude of the spectrum trace at each frequency will vary slightly with fit, but should fatigue begin to set in, or major wear, some of the frequency levels will alter significantly. For instance the rotational frequency level will rise dramatically with increased out-of-balance.

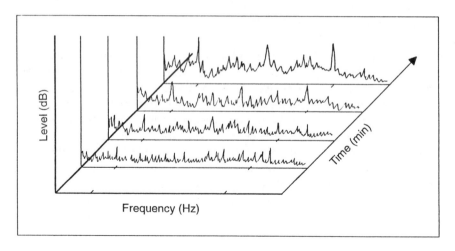

Figure 7.11 Spectrum change with time (waterfall plot).

Side-band analysis (Fig. 7.12)

Where two frequencies are affected by the same fault a trace can be produced which shows side-bands. Side-bands are where two peaks are generated at equal distance either side of the major peak. In the case of a tooth missing on a gear wheel the major peak will be the tooth meshing frequency with side-bands at the equivalent of the rotational frequency either side.

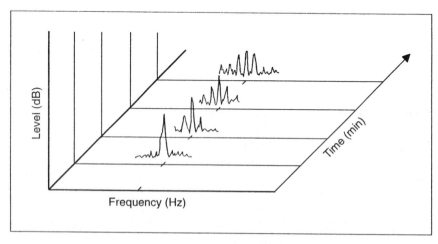

Figure 7.12 Side-bands at a component frequency

Harmonic analysis (Fig. 7.13)

While there is a fundamental frequency, e.g. the meshing frequency, a pure note is not generated. As on a stringed instrument the playing of one note actually produces many harmonics up the scale. These harmonics are at integral multiples of the fundamental frequency – two times, three times, four times etc. The same also happens with machinery; a pure tone is not generated owing to the shape and construction. If some parts have non-linear stiffness, as with a two-stage bearing mounting, subharmonics will also be generated, e.g. at one-half the fundamental frequency. A fault developing will cause the ratio of the levels of the harmonics to change.

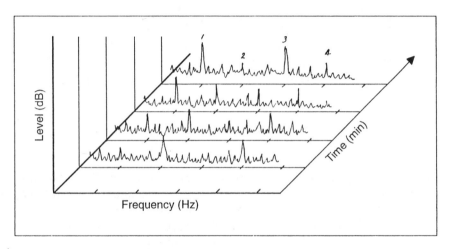

Figure 7.13 Harmonics of a deteriorating pump.

Cepstrum analysis (Fig. 7.14)

It should be apparent from the discussion of harmonic analysis that there is a distinct uniform separation of harmonics (equivalent to the fundamental frequency). If a measure were arranged to examine just the intensity of the signals which are separated by this (fundamental frequency) certain amount, then any increase in the harmonics content would be greatly magnified and would appear as just one peak. This is precisely what happens with cepstrum (pronounced 'kepstrum') analysis. In simple terms it could be defined as the FFT of the logarithmic spectrum obtained from the conventional FFT – a sort of 'FFT of an FFT'. There have been cases where none of the earlier analysis suggestions made has detected a fault developing but cepstrum analysis did.

It should be pointed out that, because of this doubling up of the transform, the first part of all the technical words in cepstrum analysis is reversed, i.e. spectrum becomes cepstrum, frequency becomes quefrency and harmonics becomes rahmonics.

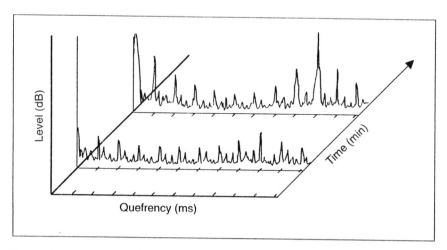

Figure 7.14 Cepstrum of a deteriorating pump.

Kurtosis

Kurtosis analysis has been restricted almost exclusively to bearings where a few specific frequency ranges are examined (e.g. 3–5 kHz, 5–10 kHz, 10–15 kHz etc.). It is a statistical analysis of the time-based (time domain) signal and looks at the fourth moment of the spectral amplitude difference from the mean level. The equation is

$$K = \frac{1}{\delta^4} \sum_{i=1}^{(N)} \frac{(x_i - \bar{x})^4}{N}$$

where δ^4 is the variance squared, N the number of samples, $\bar{x}$ the mean value of samples and x_i an individual sample.

A normal distribution has a kurtosis (K) value of 3, and this is found to be the acceptable condition of a bearing; if, however, it rises to, say, 5 it is likely that there is serious deterioration. Supposedly, the K value is not very sensitive to bearing load or running speed. However, the technique is not greatly used.

Other

There are other techniques of analysis with varying degrees of success depending on the application. One is adaptive noise cancelling which, by use of feed-back, is able to improve greatly a signal in a 'noisy' environment. Another technique is ESP – envelope signal processing – which enhances the important frequencies associated with a particular component such as a bearing. (A band-pass filter is used on the time domain signal to allow through only the frequencies which relate to the bearing, say. The FFT is then taken of this prefiltered signal.) Neural networking is beginning to be used with this to achieve a highly advanced and capable vibration analysis technique.

(c) Discussion of above vibration analysis techniques

While it is possible to calculate the frequencies generated by certain rotating parts with precise accuracy – such as the tooth meshing frequency of a gear wheel – this is not particularly common with any other components. Even bearings, which have precise diameters and numbers of balls or rollers, while they can have their frequencies of

Table 7.4 Frequencies generated by bearings (Fig. 7.15)

Bearing part generating frequency	Frequency generated (Hz)	Comments
Shaft-rotational frequency	$f = \frac{N}{60}$	Out-of-balance
Loose bearing housing	$f = \frac{N}{120}$ or $f = \frac{N}{180}$	Radial effect may appear when temperature changes
Loose machinery, e.g. at the coupling	$f = \frac{N}{30}$	See also loose housing above
Defect on ball or roller	$f = \frac{D}{d}\left[1 - \left(\frac{d}{D}\cos\beta\right)^2\right]\frac{N}{60}$	Double if both inner and outer races are impacted
Defect on inner race	$f = \left(1 + \frac{d}{D}\cos\beta\right)\frac{N}{120}\, n$	
Defect on outer race	$f = \left(1 - \frac{d}{D}\cos\beta\right)\frac{N}{120}\, n$	
Defect on cage	$f = \left(1 - \frac{d}{D}\cos\beta\right)\frac{N}{120}$	
Plain bearing oil whirl	$f \approx 0.45\frac{N}{60}$	

contact calculated (Table 7.4, Fig. 7.15), their accuracy heavily depends on the fit and slip which actually occurs; in other words, the actual vibration frequency experienced is likely to be a 'little way' away from the calculated value. In complex machinery, with several rolling elements and sliding surfaces, it is not surprising to find it difficult to differentiate between frequencies which relate to one bearing and those which relate to other features.

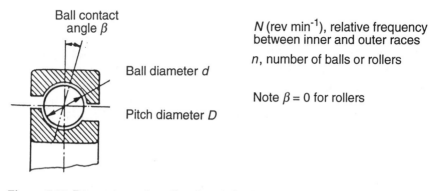

Ball contact angle β

Ball diameter d

Pitch diameter D

N (rev min^{-1}), relative frequency between inner and outer races

n, number of balls or rollers

Note $\beta = 0$ for rollers

Figure 7.15 Dimensions of a rolling bearing.

Whether a frequency relates to a rotating part (synchronous) or a stationary part (asynchronous) can be determined in variable speed applications by seeing the change in frequency with speed. Where the frequency peaks gradually rise in frequency, as the speed increases, they can be associated with a rotating part.

It is also difficult to judge what measure of change will occur with wear or fatigue. Experimental confirmation will be needed with a close watch on the actual conditions prevailing at the time. Temperature is a major issue as regards fit and clearance, and the frequencies generated on one day at one temperature may be difficult to compare with those on another day at a different temperature although the actual machine has not deteriorated at all. Another cause of change will be the loading in the system; this should be as close as possible to the same each time, if accurate comparisons are to be made.

The position of the sensors is of paramount importance. The closer the sensor is to the part likely to deteriorate, the better will be the signal-to-noise level – 'noise' in this case being from other machine parts not far away. Thick solid bearing mounts may be good for refined running, but they make vibration analysis hard work.

7.2.5 How to start

(a) Portable vibration analysers

There is a very large number of portable, hand-held, vibration analysers on the market. It is common practice for one manufacturer to have several outlets of the same instrument but with different logos attached, so there are not as many different models as would appear at first sight. However, the range is still large and requires careful thought before a purchase is made. The best arrangement is where the supplier is willing to loan an instrument for the user to test on his or her machine (and to compare with other portables). If a fault can be induced on the machine, an even better likelihood of choosing the best device will be achieved. The portable should also have a range of connecting probes suitable for a variety of applications. Fig. 7.16 shows a typical portable data collector and vibration analyser.

Portable monitors could be compared by looking at the features listed below:

- frequency range (e.g. 1 Hz to 25 kHZ);
- frequency bands for analysis of trends;
- amplitude ranges (more for greater accuracy);
- input sensors acceptable (e.g. accelerometer, velocity, displacement, temperature);
- output (e.g. RS232, baud rate);
- storage capacity (e.g. 1 Mbyte = 1024 kbytes);
- microprocessor (e.g. 16-bit data bus);
- resolution (e.g. 100 Hz, 200 Hz, etc.);
- ambient conditions acceptable (temperature and humidity);
- analysis techniques (e.g. RMS, FFT, enveloping, time averaging to 9999 samples);
- analysis time;
- keyboard;
- display;
- print out;
- alarms;
- weight;
- hours per charge;
- cost.

Figure 7.16 A portable machine vibration analyser (Diagnostic Instruments).

Some portables are dedicated to a particular function, such as bearings. In that case they will be much simpler and less expensive; some are no larger than a thick highlighter pen. Each year they will advance in sophistication as the microprocessor chips improve but they should remain very simple to use. They are often delivered with touch probes which require merely a firm hand at the same place each time the machinery is investigated.

A more obvious difference between the portables is what they actually purport to do. Some will display the vibration waveforms, most will perform FFT analysis but the storage will be limited. There will be a need to compare previous results with the current waveform. Read-out can vary from almost meaningless figures (unless you are highly skilled in the understanding of manuals) to simple coloured lights (presumably green means OK). Hence the portable should relate not only to the machine being monitored but also to the person responsible for the analysis.

A caution must also be mentioned where portables are being used for low speed machinery. The bottom end of the monitor's frequency range may be above the actual running frequency. For instance a machine running at $240 \, \mathrm{rev \, min^{-1}}$ has a running frequency of only 4 Hz so to use a monitor with a minimum of 10 Hz will miss the once-per-revolution signal.

(b) Sensors to be attached to computers

In this case it would be usual to employ a vibration specialist or sensor company to give advice and to provide the necessary connections and fittings as well as giving advice on the electronics necessary. A sensor in an inappropriate place can be totally useless. However, the arrangement, once it has been correctly fitted, will be much more reliable and repeatable than a portable. The initial cost will be higher but the running cost will be less, as the operator hours are fewer.

7.3 ULTRASONICS

7.3.1 Introduction

There are many audio frequency examples where the human ear is able to detect a mechanically induced noise when an extreme failure has occurred, such as the hiss of a steam leak, or the knocking of a big-end bearing. However, when we make attempts to detect the early stages of a failure condition our success rate is somewhat reduced; this is due to extraneous noise from other features, particularly when the noise is random.

The frequencies generated by various machine movements are not restricted to the audio range (20 Hz to 20 kHz); the generation spans many decades reaching as high as hundreds of megahertz. This is where there is a bonus – the use of the higher frequencies for detection (above 20 kHz) means that there is a much clearer spectrum with less confusion from general background noise.

The frequency band up to 20 kHz or perhaps 25 kHz has been covered in the previous section on vibration. This section now looks at the different techniques used above those frequencies where the descriptions 'ultrasonics', 'shock pulse', 'acoustic emission' and 'stress wave' are used. Although there is an overlap between these terms, it is helpful in the condition monitoring field to consider them as follows:

- ultrasonics, 20 kHz–100 kHz, including
- shock pulse, 32 kHz;
- acoustic emission, 100 kHz–1 MHz, including
- stress wave, 100 kHz–600 kHz.

7.3.2 Brief description

All movement within a material structure generates stress waves as energy is released. Actions associated with deterioration of machines, such as impacts, fatigue, friction, turbulence and material loss, all produce 'sounds' over a wide range of frequency. However, these sounds are not random; there are clearly defined 'signatures' associated with different processes.

The ultrasonic monitoring (Fig. 7.17) filters out the audio sounds (and other unwanted frequencies) and it is able to sense the stress waves generated from specific faults. Most ultrasonic machine monitoring is conducted with solid-borne (or liquid-borne) transmission of the energy waves, which means that each machine is usually isolated from any influence of adjacent machinery. (Airborne ultrasonics is used where direct contact is not feasible, or where transmission through a crack or leak is being investigated.)

The indication of the ultrasonic presence may be displaced visually (on a VDU), audibly (at a modulated frequency) or electrically to an electronic monitor or computer.

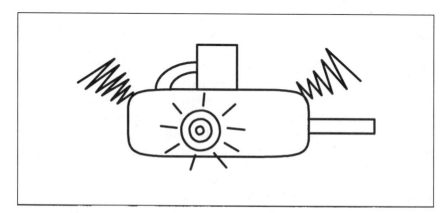

Figure 7.17 Ultrasonic monitoring.

7.3.3 Practical applications

The following is a selection of the many possible failure mode features which can be detected:

- bearing wear or fatigue (plain, roller and ball type bearings);
- valve and valve seat wear;
- cavitation in pumps;
- diesel engine injector nozzle wear;
- crack propagation;
- gas burner jet performance;
- pipe suction leaks;
- container leaks (including welds);
- internal component leaks;
- underground water leaks;
- source of mechanical noise;
- arcing in electrical junction boxes or switches;
- electrical insulation breakdown (down to 45 pC);
- static discharge.

In other words, failures can be detected wherever there are small orifices (to allow change of flow if they enlarge), wherever there are sharp edges which may change in shape and wherever there are high forces exerted of a macroscopic size.

Holroyd Instruments have produced a list of reasons why the energy level may rise or may fall when a fault develops in a pump (Table 7.5).

Table 7.5 Factors affecting the ultrasonic level from an operating pump (assuming the same pump speed)

Factors causing an increase	*Factors causing a decrease*
Bearing damage	Liquid starvation (i.e. air pumped)
Loss of bearing lubrication	Complete pump failure
Cavitation in the fluid	

7.3.4 Detailed discussion of ultrasonic and acoustic emission monitoring

There are a number of ultrasonic detectors of various names. Some are tuned to one frequency or a frequency band, others scan across a range to find the most important frequency for the application and another can be set at the most suitable frequency band over a very wide range. (The transducer used has a maximum sensitivity at its resonant frequency.) The manufacturers of single-frequency detectors claim that their frequency is the main one for the specific application such as bearings. The scanned frequency devices may have the edge here if there are doubts about what is suitable for a new application, particularly if there are several components to be monitored; however, check the sensitivity.

Table 7.5 indicates that just looking at a fixed level, if there are more than one or two features involved, may not be adequate. On the other hand if the only problem is likely to be bearing spalls or fatigue, then a single frequency will be preferable.

Ultrasonic monitoring, at least in the acoustic emission band, tends to complement vibration monitoring; both have their advantages and disadvantages and they are not the same. Table 7.6, again from Holroyd Instruments, gives some of these comparisons.

Table 7.6 The comparison of acoustic emission and vibration monitoring

Feature	Acoustic emission	Vibration
Detection	Omnidirectional	Unidirectional
Propagation	Weak resonances	Strong resonances
Standing waves	Unimportant	Important
Signal sensitivity	High	Low
Noise sensitivity	Low	High
Slow machinery	Easy to apply	Difficult to apply

It could be added that the lower ultrasonic frequencies, e.g. 32 kHz, detect the bearing response better than vibration analysers, but vibration analysers pick up the general vibration levels of the machine which the ultrasonics miss.

One of the problems with lubricated and fluid power systems is the damping which occurs to the low frequency signals. It may be wondered whether the high frequencies, that we are now discussing, are also damped: apparently not, even the material damping, which could occur, only arises about 20 MHz, well clear of the 600kHz maximum normally used.

Although a direct contact with the hardware of the system is usual, it is not essential. By using a directional reflector it is possible to detect ultrasonic action (and impending faults) at many metres distance.

The four types described in the following will help to set the scene.

(a) Shock pulse meter and Detectaids

The shock pulse meter (SPM) has been used for bearings since the early 1970s. The shock pulses emitted by rolling element bearings are picked up by a shock pulse transducer mounted, or held, on the bearing housing. The transducer is tuned to 32 kHz, which is sufficiently high to avoid the majority of the unwanted frequencies. (For leak detection a band from 25 kHz to 40 kHz is covered with a detection sensitivity as low as 1 nbar.) The bearing's shock pulse level (decibel shock value dB_{sv}) is a function of its size, rotational speed and operating condition (Fig.7.18).

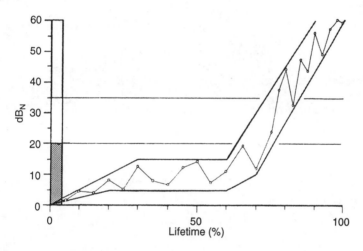

Figure 7.18 Typical SPM readings over the life of a bearing.

Detectaids (UK) Ltd use a single frequency close to 40 kHz. Amongst other applications this has been successful in the monitoring of valves, in particular seat wear; fluid leaks as low as a few parts per million for both gaseous and liquid systems have also been detected.

(b) Senaco Plus stress wave sensor

The broad-band sensor covers the frequency range from 100 kHz to 600 kHz. This is a solid-borne detector connected with a single bolt as close as possible to the component or fluid being monitored (Fig. 7.19). This sensor, like other ultrasonic sensors, is able to detect flow turbulence and cavitation as well as friction and defect growth, and is hence also sold as a non-invasive flow device. With the supplied electronic control the appropriate set-up can be arranged.

Figure 7.19 The Stresswave Senaco Plus sensor (Milltronics).

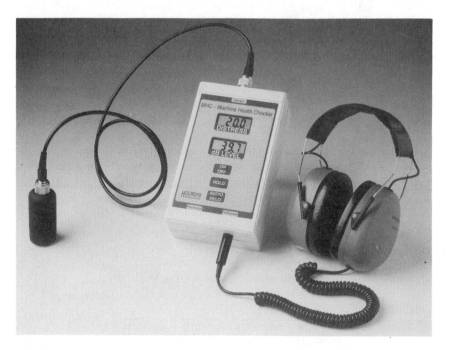

Figure 7.20 The portable Machine Health Checker (Holroyd Instruments).

(c) Machine health checker

Slow speed rolling bearings and plain bearings can be monitored at the 100 kHz chosen frequency. This is the acoustic emission region which is used in the machine health checker. Such a high frequency will also detect many of the other examples mentioned above. Both portable (Fig. 7.20) and hardwired systems are available. The enveloping involved is undertaken by rectifying and averaging the signal.

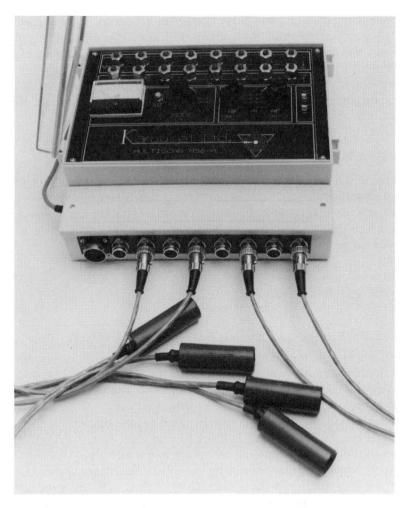

Figure 7.21 The Karousel multi-scan ultrasonic recorder (Detectaids).

(d) Ultraprobe

This hand-held device has a continuously variable frequency range between 20 kHz and 100 kHz in order to determine the best frequency for the application. It is an airborne ultrasonic unit, although a solid probe can be fitted. The frequency 'tuning' enables the user to tune into the sounds associated with the particular problem and possibly to lock onto that feature, or to note a change in frequency pattern associated with wear characteristics.

(e) Multirecording

The recording of ultrasonic signals may be undertaken by electronic multichannel data acquisition units. With these a considerable number (perhaps 64) of independent ultrasonic sensors can be simultaneously monitored. By using pre-amplifiers on the sensors up to 100 m remote monitoring is possible. The particular unit developed by Detectaids has an RF response of ±3 dB from 38 kHz to 42 kHz in order to cover the basic ultrasonic frequency of 40 kHz (Fig. 7.21).

7.3.5 How to start

Because of the very different responses that can result depending on the type of machinery, it is best to arrange for a demonstration from several of the suppliers. It would be helpful to consider what type of component or system is going to be reasonably monitored in this way – check with the various examples given in this section 7.3.

7.4 WEAR DEBRIS ANALYSIS

7.4.1 Introduction

The oil in a machine probably holds more information about the condition of the machine than any other feature. That is not to say that all the other techniques are irrelevant; they certainly are not. The oil content cannot detect everything, and there are some features which may be clearer with another method, but the oil has a wider band of possibilities.

The oil itself will be discussed in section 7.5. The present section looks at the solid debris and contaminant found in the oil.

'Wear debris' is strictly speaking the solid debris removed from the material construction within a machine – through the process of wear. Wear can occur as a result of adhesion, abrasion, fatigue and tribochemistry. However, it is not the purpose of this book to look at these wear processes but rather to sense how the technique of wear debris analysis can detect and analyse this debris with considerable success.

The 'debris' in the oil – the foreign particulate – is not confined to wear debris. Depending on the situation, there will be a considerable amount of debris ingested from the atmosphere or remaining from the build of the machine; in this case this debris is called 'contamination' in the sense that it will cause trouble. This means that debris analysis not only detects the machine beginning to deteriorate, but also detects the harmful contaminants which will cause the machine to deteriorate. That is one step ahead, which is virtually proactive monitoring. Table 7.7 gives a selection of shapes which may typically be found in an investigation of a machine oil.

Table 7.7 Debris particles found in a typical machine oil

Particle shape	Typical names	Some possible origins
	Spheres	Metal fatigue Welding 'sparks' Glass peening beads
	Pebbles and smooth ovoids	Quarry dust Atmospheric dust
	Chunks and slabs	Metal fatigue Bearing pitting Rock debris
	Platelets and flakes	Running-in metal wear Paint or rust Copper in grease
	Curls, spirals and slivers	Machining debris produced at high temperature
	Rolls	Probably similar to platelets but in a rolled form
	Strands and fibres	Polymers Cotton and wood fibres Occasionally metal

Another point to note is that the monitoring may occur in line, on line or off line as shown in Fig. 7.22. There are advantages and disadvantages with each type as outlined in Table 7.8.

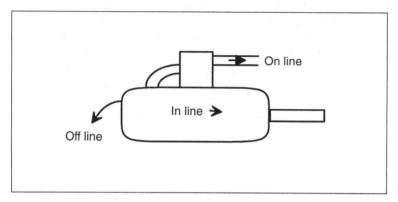

Figure 7.22 Wear debris monitoring.

Table 7.8 Different sampling and analysis positions

Type	Description	Advantages	Disadvantages
In-line	All the oil that passes is tested by the monitor	Nothing is missed No outside influence Immediate result	Difficult to fit Can restrict flow May not be 100% Does not keep evidence
On-line	A proportion of oil is sampled and tested by direct connection and returned or ejected to another container	No outside influence Easier to fit Little influence on flow Immediate result	If proportion is small, it may not be fully representative of the system
Off-line	A proportion is taken away for analysis	Cheap for one-off	Badly influenced by the environment during sampling and transport Messy

7.4.2 Brief description

Wear debris analysis is that part of oil analysis which detects foreign particles and contaminant in the fluid. From the analysis of the particles is it possible to determine wear and deterioration of machinery at a very early stage. 'Contaminant' analysis indicates what may eventually cause wear and failure if no maintenance is applied.

7.4.3 Practical applications

All oil-lubricated and fluid power systems are applicable. However, not all parts would be appropriate because the liquid must pass over the critical component, i.e. the component must be oil wetted.

Particular applications are those where the failures are likely to be due to wear or fatigue. Bearings and gears are frequent culprits of wear, such as pitting due to overloading. Pitting is the result of fatigue originating below the surface, cracks developing and small pieces of metal becoming released. If the early stage of such wear can be detected then replacement can be actioned before any consequential damage results. Surface scuffing is another culprit where excessively high shock loads, or a dry start-up, cause surface wear on a gear tooth before a film of lubricant can be applied.

In fluid power systems, as well as the more conventional mechanical wear failures experienced by valves and pumps, cavitation also causes debris to be shed. Cavitation is due to a low pressure at the inlet to a pump causing local vaporization of the fluid under vacuum conditions; tiny air bubbles are released and as they collapse on the metal surfaces extremely high pressures are generated; the implosions cause the surfaces to be eroded. This is also discussed in section 7.8.1.

7.4.4 Detailed discussion of wear debris monitoring

There are three critical parts in the monitoring programme:

1. attachment (and sampling);
2. detection;
3. analysis.

(a) Attachment

There needs to be a means of extracting oil from the system, or of getting it to a detector in the line. In other words, an insertion needs to be made into the flow or reservoir. This must be treated with the same care as the detection and analysis; if the sample is unrepresentative of what is actually in the fluid, then no matter how good the analysis is it cannot give the true result.

Some examples of points of attachment are indicated in Fig. 7.23. Note that they are where the fluid is thoroughly being mixed up (after corners), they are in the flow itself and not in any dead ends or dead regions of the reservoir (well up in the fluid). They are also, in the case of horizontal pipes, attached to the side rather than in the bottom (where debris may over-collect) or on the top (where air may confuse if the pipe is not full) (Fig. 2.4).

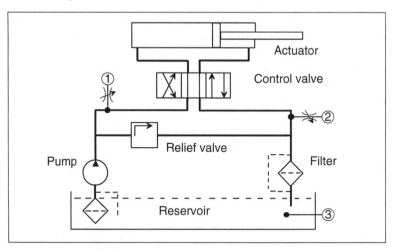

Figure 7.23 Three turbulent positions for sampling fluids for debris and contaminant.

If a bottle is used to take a sample, then it needs to be cleaned by a laboratory able to quote the cleanliness level to which it is cleaned, e.g. 'less than 500 particles >5 μm per bottle'. Note that sterilization is not the same – that merely kills the debris but does not remove it. The bottle should have a non-shedding screw cap or a special seal.

The time at which samples are taken is also critical. The fluid should be flowing, yes, but it should also have reached the operating temperature and be thoroughly mixed. A 30 min warm-up period is quite common. When trending is used (section 10.1) it is particularly important to test the oil at exactly the same conditions each time.

(b) Detection

The detection is tied up in some ways with the analysis. If we just want to know the quantity of debris which is in the system fluid, then the detection can be relatively simple, such as a gravimetric method (i.e. weighing the debris collected on a filter when, say, 100 ml of fluid has passed through). However, at the other extreme, if we want to know precisely which part is wearing then a much more complex and expensive technique will be called for, such as nuclear seeding or spectrometry.

(Note that nuclear seeding requires the part to have been exposed previously to radiation, and then a Geiger counter detects either the particles released or the lower level of radiation on the part due to wear.)

There is also the problem as to how to define a particle or quantity of particles. There is no such thing as a single dimension size of a complex particle (except for a sphere or regular polyhedron); we, therefore, need to know why a 'size' is required. Is it to relate to an international absolute level or just to trend any growth of particle increase? Is it rather to sense how badly the system will be worn by the contaminant (use a 'wear rate' monitor) or to sense how the particles block the system (use a 'blocking' monitor)? A two-dimensional optical dimension based on the cross-sectional area has little relevance in engineering wear but it is the usual dimension used in international standards.

The following are a selection of detection (and analysis) techniques used in monitors.

Electrical sensing zone (Coulter) (Fig. 7.24)

This technique is unusual in that it is able to determine the particle 'size' in terms of the true volume of the particle. As each particle passes through a small orifice or tube, the electrical resistance between the two ends of the liquid column in the tube is changed. The technique will only work for electrically conducting liquids, but most particle types are able to be detected. It is currently only used as a laboratory instrument.

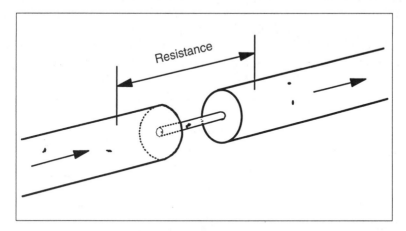

Figure 7.24 ESZ technique.

Electroacoustic (Fig.7.25)

An alternating electrical field, at varying frequency (in the megahertz band), is applied to the volume of liquid. By measuring the magnitude and phase angle of the sound wave produced by the particles, it is possible to determine both the particle size and the effective charge (zeta potential). A large range of particle concentration is possible from 0.5% to over 40%, but the particles must have a charge. No dilution is necessary.

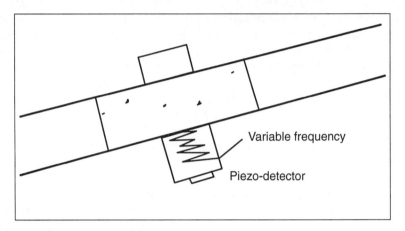

Figure 7.25 Electroacoustic technique.

Filter blockage (mesh obscuration) (Fig. 7.26)

This technique can be used both on-line and off-line with no restriction on fluid mixes or particle types. By merely sensing the change in flow characteristics (flow and pressure drop), the level of particle count associated with the pore size of the filter can be determined. Continuous operation occurs when the flow is reversed and the filter back-flushed. Several refined filters can be fitted and a particle distribution obtained.

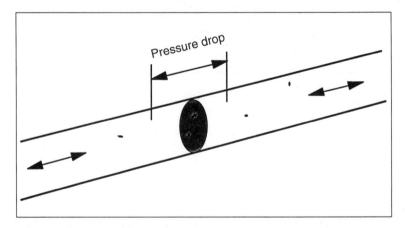

Figure 7.26 Filter blockage techniques.

Inductance (Fig. 7.27)

Any metal particles passing through a coil cause a change in the inductance. The size of individual particles which can be detected is limited by the sensitivity of the electronics and the shielding from electronic 'noise'. Ferrous and non-ferrous particles can be detected separately. Air bubbles can be isolated. There is no disturbance to the flow, but a small bore size has better sensitivity.

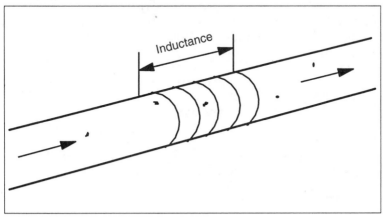

Figure 7.27 Inductance technique.

Magnetic attraction (Fig. 7.28)

There are numerous devices made which may either trap particles, magnetically or otherwise, and sense them statically or sense them as they pass, by detecting a change in magnetic flux. Only the presence of

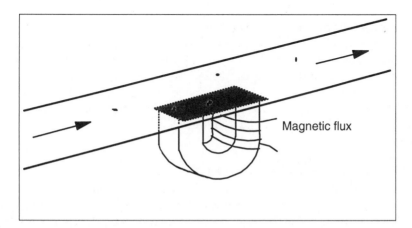

Figure 7.28 Magnetic technique.

ferromagnetic particles can be accurately determined. The devices do not normally give a particle count, but rather a total view of the 'large' and 'small' particulate present. Some monitors work continuously.

Optical–Fraunhofer (Fig. 7.29)

There are many instruments based on the Fraunhofer detection method. They are designed to obtain a very accurate assessment of the particle distribution from 0.1 μm upwards. Some instruments can detect even smaller particles. An array of detectors ahead of the beam receives the light rays, diffracted according to the size of the particles – the smaller the particle the greater the diffraction.

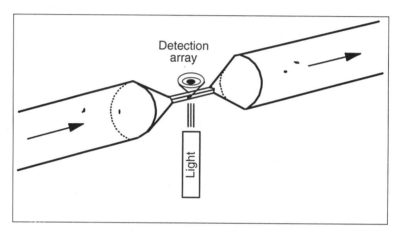

Figure 7.29 Fraunhofer technique.

Optical–light obscuration (Fig. 7.30)

Light obscuration is the simplest type of light instrument, looking solely at the shadows cast when particles pass through the light beam. Much work has been undertaken to reduce the probability of more than one particle being counted at any one time (counted as one larger particle). Although originally only a laboratory instrument, more devices are becoming available for field use in the 'cleaner' industries such as fluid power.

Optical–time of transition (Fig. 7.31)

This technique is relatively new but has become popular owing to its ability to examine concentrations of particulate far beyond the capabilities of other optical devices – even as high as 70% concentration of particulate. The rotating focused light beam crosses over particles

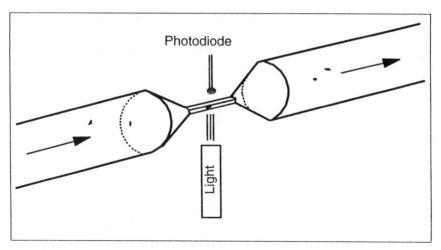

Figure 7.30 Light obscuration technique.

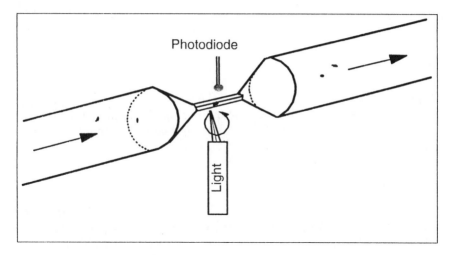

Figure 7.31 Time of transition technique.

and the time of transition is measured – longer for larger particles. It is possible to put a probe into the liquid or to pump the liquid through the instrument.

Wear of thin film (Fulmer) (Fig. 7.32)
This is a very specific device which can look at the concentrations of particles which wear away the thin conducting plate used as the sensor. Provided that the viscosity of the fluid and the type of particle are

known (e.g. hardness and sharpness), the device can be accurately calibrated and sense very small changes in particle levels. Any metal or relatively hard particle can be detected, and the device can be continuous. A PC plug-in version is available.

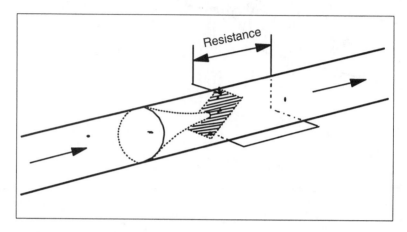

Figure 7.32 Fulmer technique.

Some of the above techniques are covered by the British Standard BS 3406. A typical on-line analyser is shown in Fig. 7.33. This is based on the filter blockage technique; the portable unit is shown connected to a hydraulic system.

(c) Analysis

A major technique used off-line is that involving the passing of a fixed amount of fluid through a membrane. The arrangement is shown in Fig. 7.34. While the microscope examination of debris, such as in Fig. 7.34, will give a visual impression of the debris (Table 7.7) which can be identified in a wear atlas, it is a little subjective and is unable to give elemental information. (Incidentally, the analysis of the membrane can be performed more rapidly and less subjectively using image analysis with a video camera observing through the microscope eye piece.)

Considerable off-line analysis is also conducted with ferrography – based on magnetic attraction. While the PCM (Fig. 7.33) looks at all particulate debris in the oil, the Ferrograph is primarily a detector and analyser of magnetic and paramagnetic metals. The idea of the ferrographic technique is to deposit on a slide the magnetic debris in strands of varying particle dimension – the larger particles being deposited first,

Figure 7.33 The Pall Cleanliness Monitor (PCM) based on filter blockage (Pall).

owing to a weaker magnetic field, and the smaller silt particles last. In this way not only is the debris well spaced out, but an immediate appreciation can be made of whether fatigue debris or normal wear silt is present. Currently there are two types of slide available, the 60 mm straight slide from the older Ferrograph and the circular ring deposit slide (from the rotary particle depositor (RPD)); these are shown diagrammatically in Fig. 7.35, with an example of the appearance under a microscope in Fig. 7.36. A photograph of the RPD is shown in Fig. 7.37.

Figure 7.34 The detection and analysis of debris using a membrane (Millipore).

Mention has already been made of 'atlases'. These are compilations of real debris photographs (such as Fig. 7.36) taken from membrane samples or ferrographic slides. These allow the engineer to identify rapidly particular particles and hence determine their origin. Unfortunately it is not quite as easy as that because particles vary so much in shape; however, with experience the atlas is an invaluable source of visual information. Three atlases are mentioned in the Bibliography.

Another very important and well-used off-line technique is that of spectroscopy, sometimes called SOA or SOAP – spectrometric oil analysis programme. This is able to identify the elements present in an oil or particle. Because this relates more to 'oil analysis' it is described in detail in section 7.5.

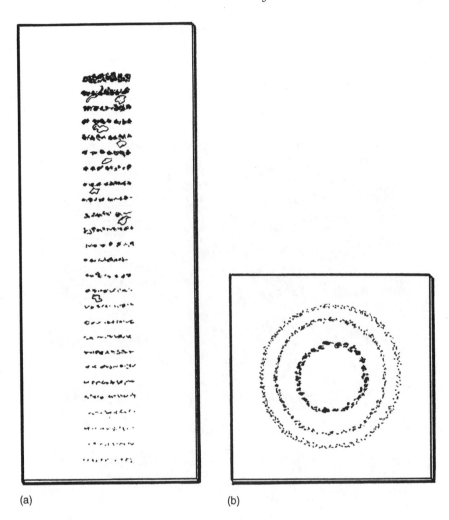

Figure 7.35 The two types of ferrographic slides: (a) from the Ferrograph; (b) from the RPD.

Any of the monitors described earlier can be used in different ways:

1. to assess the condition of the machinery from an examination of

   - the total particulate present,
   - the metal particulate present and
   - the ferrous particulate present

   in terms of size range, size distribution and large-to-small particle ratio;

2. to assess the seriousness of the contamination

  • likely to produce failure,
  • likely to produce wear and
  • likely to produce blocking

in terms of contamination classification and contamination severity index.

Figure 7.36 A typical separation of particles seen under a microscope.

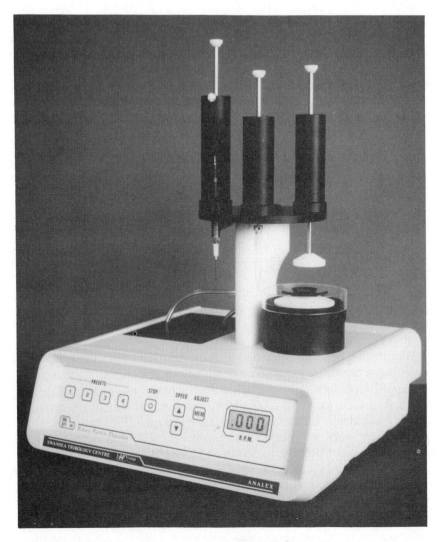

Figure 7.37 The Rotary Particle Depositor (Swansea).

While there are some absolute standards (such as NAS 1638 and ISO 4406 in Chapter 14), most condition monitoring will be trend monitoring. Chapter 10 goes into detail on both these accounts.

7.4.5 How to start

Wear debris analysis is a type of monitoring in which the engineer can gradually learn and gradually improve the level of instrumentation. For instance, it is best to start with the off-line membrane sampling, for which we need only to purchase a sampler (section 9.4), the Millipore glassware and vacuum pump, some membrane filters (0.8μm) and a microscope (Fig. 7.34). There will be a need to examine one of the wear atlases (section 10.2 and the Bibliography).

The progression from there will depend on the debris. If there is a considerable amount of ingested debris from the atmosphere – as in quarrying – then a monitor which detects all particulate will be essential. If the debris is just wear debris – then a metallic sensing monitor is required. If the debris is only ferrous then the magnetic types would be acceptable.

7.5 OIL ANALYSIS

7.5.1 Introduction

Oil enables machinery to run 'smoothly'. It provides cooling. It removes contaminant. It prevents corrosion and, in fluid power applications, it is the channel of power. It is no wonder that within the oil there is much evidence of condition.

Monitoring of oil in the past has been primarily with the purpose of deciding when the oil should be changed, i.e. to determine whether the oil itself had deteriorated to such a degree that it no longer could fulfil its function. However, the oil picks up evidence as it flows around a system – evidence of the condition of the machine. This bears considerable resemblance to human deteriorating conditions which can be detected from the blood circulation and sampling.

The wear debris (and solid particulate contaminant) has already been discussed in the previous section. The current section will now look at the other indicators featured in the oil.

7.5.2 Brief description (Fig. 7.38)

By examining the characteristics of an oil in use, various indications can be detected of the mechanical system which may be developing a fault. The oil itself may also be deteriorating but at an increased rate owing to the machine fault.

7.5.3 Practical applications

All oil-lubricated systems are applicable as well as hydraulic (fluid power) systems. This includes grease-lubricated items such as bearings. However, as a monitoring system in its own right, rather than an adjunct to other monitoring, oil monitoring is most appropriate where other techniques are difficult to fit or where initial costs have to be kept to a minimum.

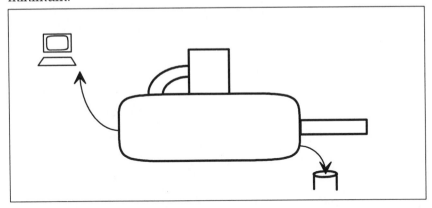

Figure 7.38 Oil analysis.

7.5.4 Detailed discussion of oil analysis techniques

Before discussing the various features which can be detected from oil monitoring it is important to appreciate the properties of oils. The list in Table 7.9 includes the major basic properties usually provided by oil suppliers. Many oils, however, are dedicated to specific functions, such as gear oils, or hydraulic oils, so they will have a concentration of more specific additives as indicated in Table 7.10.

The monitoring of additives could well provide valuable indications of how the machine has been operating where certain additives have been consumed too rapidly, such as anti-wear or viscosity improvers. However, apart from the oil manufacturers and their associated laboratories, it is not easy to acquire the necessary background information in sufficient detail on the additives – they are company confidential. The monitoring of the other properties is not a problem, and may be just as beneficial.

As regards machine monitoring, the list of properties in Table 7.9 might seem a little obscure, and certainly that would be the case if such were used solely to detect, say, the leakage of water into the oil due to a fracture or loosened fitting (there are specific water-in-oil monitoring techniques as described later in this chapter). Solid debris could also be detected, particularly with the viscosity and density testing. However, there are many other possibles as shown in Table 7.11 both for a rise and for a fall in the specific values.

Table 7.9 Oil properties which can be monitored

Property	Units	Description	Comments
Viscosity (kinematic)	mm^2s^{-1}	A measure of the oil's resistance to flow	Oil viscosity drops substantially with rise in temperature
Viscosity index (VI)		A measure of the oil's resistance to dropping in viscosity	From 0 to 300; the higher the value the less change of viscosity with temperature rise
Density (ρ)	$kg\,m^{-3}$	A measure of the oil's mass per unit volume	Typical oil would be from $880\,kg\,m^{-3}$ (20°C) to $830\,kg\,m^{-3}$ (100°C), varying with pressure
Total acid number (TAN)	mg $KOH\,g^{-1}$	Amount of potassium hydroxide neutralizing 1 g acid sample	Increases with oxidation and in the presence of high sulphur diesel fuels
Total base number (TBN)	mg $KOH\,g^{-1}$	Acid equivalent to KOH needed to neutralize 1g base	Included to restrict acids in their corrosive effect
Water content	ppm	Dissolved, but at higher levels may form a fine dispersion of droplets	Unhelpful both for the power fluid and for the lubricant (even at 100ppm)
Pour point	°C	Lowest temperature at which the oil will just pour from a container	Oils are normally used at least 10°C above the pour point
Flash point	°C	Temperature at which vapours given off ignite in presence of a flame	Typically between 150°C and 250°C for a mineral oil

Table 7.10 Oil additives

Additive	Comments
Viscosity index improver	Long chain polymers which tend to straighten out with increase in temperature are sheared apart by mechanical action.
Rust inhibitor	A compound with high polar attraction to metal provides a protective surface, but it may attack non-ferrous metals
Corrosion inhibitor	Organic acids will need counteracting by a high alkalinity in the oil, i.e. a high TBN; important where water leakage is excessive
Anti-wear	Mild wear is reduced by an adsorbed film on the metal surface
Extreme pressure agent	High pressure causes a surface chemical reaction
Anti-oxidation	Reduces oxidation by the formation of inactive compounds
Anti-foam	Defoamant containing silicone discourages foam but may also discourage the release of entrained air into the atmosphere
Pour point depressant	Inhibits formation of wax crystals; can lower pour point by 20°C
Demulsifier	Encourages the water present to separate from the oil
Emulsifier	Opposite to demulsifier – applications require an emulsion to form
Bactericide	A biocide controls the growth of bacteria and fungi

Table 7.11 Fluid changes

Fluid feature	Change	Reason	Comments
Debris	Increased level	General wear increasing	Section 7.4
	Specific elements	Specific wear	Table 7.12
Colour	Turbidity rise	General wear	Section 7.1.4a
	Change in colour	Leakage, corrosion	When two fluids mix, there could well be a colour change Corrosion particles can cause a colour change
Temperature	Rise	Power losses	Heat generated in friction
	Drop	Internal leakage	Although leakage can cause a rise in temperature, initially there is likely to be a drop
Viscosity	Rise	Corrosion	Solid particles will raise the viscosity
		Process liquid leakage	Viscosity will rise if the fluid leaking in is of greater viscosity
		Chemical action	A sludge may be produced Volatiles may evaporate in the oil
	Drop	Fuel dilution Coolant leakage Additive losses	Fuel or coolant leaking into the oil will reduce the viscosity Viscosity improvers can be destroyed by excessive mechanical action
Density	Rise	Chemical action Mixed fluids	Density is less dependent on temperature than is viscosity Oxidation may occur owing to overheating and/or presence of air
	Drop	Fuel dilution	
Water content	Rise	Leakage	Initial, as-new, concentration may be 100 ppm which could be acceptable Higher levels cause corrosion, viscosity drop and deposit formation
	Drop	Efficient water filtration	
TBN	Drop	Water content High temperatures Oxidation	Too low a level will result in corrosion, wear and formation of deposits
Elements	Rise in concentration	Debris addition	Wear or solid contamination
	Drop in concentration	Filtration	

7.5.5 How to start

There are a number of instruments which may be able to monitor the features mentioned in Table 7.11.

(a) Debris

This is discussed in section 7.4.

(b) Colour

This is discussed in section 7.1.

(c) Temperature

This is discussed in section 7.9.

(d) Viscosity

Viscosity is officially measured by complex glasswork in a temperature-controlled bath; however, the value can be determined from falling balls or a rotating paddle torque or a vibrating 'tuning fork'. In some of the latter techniques it is possible to measure viscosity on-line while the system is running and to include density and temperature. Viscometers are sold in a number of designs; one example, that of the vibrating sensor type, is shown in Fig. 7.39.

The density of the fluid is determined by measuring the resonant frequency. The damping conveys the viscosity from

$$\text{Viscosity} = \left(\frac{\text{bandwith}}{\text{resonant frequency}} \right)^2$$

where the bandwith is the frequency between the -3 dB points either side of the resonance.

(e) Density

Densitometers are basically the weighing of a known volume; however, a tuning fork technique is commonly used with a suitable calibration (also includes the need to monitor temperature). A multi-device is not uncommon as in Fig. 7.39.

Figure 7.39 A 7827 vibrating sensor viscometer (Solartron Transducers).

(f) Water content

Water content is measured in a variety of ways, some more appropriate for very low ppm and others for greater percentages (Table 7.12).

The percentages given are those quoted by some of the suppliers of the instruments at the current time, and the accuracy and repeatability will vary with instrument design and construction. Oil laboratories have tended to use titration as their means of assessing water in hydraulic oil samples; for larger percentages in lubricating oil it has been common to use Dean and Stark distillation.

Another problem with some of the analysers is that the calibration depends on using a 'water-free' oil of the same type as is being tested. This may not be easy to obtain, and even less easy to retain. In practice, for the higher accuracies, it will be necessary to have the calibration fluid checked by possibly titration or distillation. If an absolute value is not essential, i.e. the monitoring is of a trend, then great care must be taken to ensure the calibration liquid is maintained at the same level of water (sealed containers when not in use).

Conversely, oil on water can be an indicator of machine deterioration. For this there are also monitors available. Some of the water-in-oil

Table 7.12 Water-in-oil monitors

Technique	Water range %	Comments
Chemical reaction	0.05–100	A pressure is caused to build up by the chemical reaction when calcium carbide (or hydride) is mixed with the oil–water
Coulometric titration	0.0001–100	This is normally used in the laboratory but because of its portability can be used in the field
Crackle test	Perhaps 0.05+	If the oil–water is placed on a hot 'plate' there will be a 'crackle' if water is present
Dean and Stark	0.1–100	More suited for laboratory work for the larger percentages
Densitometer	See comments	The greater the accuracy of weighing the oil–water, the lower the percentage of water detectable; 5 decimal places preferred
Dielectric constant–capacitance	0.01–10 (possibly 50)	Temperature, density and particulate levels need to be controlled or compensated for in order to achieve the lower levels
Electromagnetic wave absorption	0–100	Similar to the microwave oven, electromagnetic waves are absorbed differently by water and by oils, water rising in temperature more quickly
Infra-red absorption	0.005–0.5 (possibly to 3%)	Two frequencies are used, one which water absorbs and the other, reference, which is not changed by water; detectors measure the change in transmission of the infra-red beams
Infra-red reflection	0.5–90	As IR absorption but with the reflected energy measured
Steam generation	0.1–3	A small sample is heated and the pressure achieved indicates the proportion of water present
Visi-cross	0.05+	The cloudiness of an otherwise clear hydraulic fluid prevents a cross being seen as it is immersed into the fluid

monitors given in Table 7.12, of suitable percentage, can be used, but there are also specific oil-on-water monitors; in this case a light beam projected onto the water senses any change in the reflectance character-istics – detecting as little as a 0.1µm thickness of oil well before it is visible to the naked eye. Some other devices use the scatter of light at different angles due to the presence of oil in the water. Synthetic oils have a natural fluorescence and can be detected by using a fluorescence meter.

Various individual property test monitors are available such as TBN and TAN monitors, or simple on-site analysis kits. More advanced

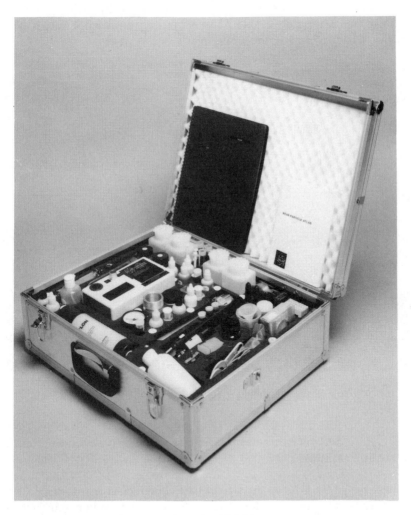

Figure 7.40 A comprehensive oil test kit (Oilab).

portable kits may contain simple tests to cover such features as debris, viscosity, TAN, TBN, water content. One example is the Oilab kit shown in Fig. 7.40.

(g) Element content – spectroscopy

The measurement of the elements in an oil requires complex and advanced instrumentation, usually of considerable cost (from £7,000 to say £100,000 or even over £1,000,000 in one case) – Figure 7.41 illustrates an inductively coupled plasma (ICP) spectrometer (Spectro). Particles in the oil, or the oil itself, can be analysed by such spectrometers to determine their elemental content. This is invaluable for identifying the true origin of the particulate. Table 7.13 lays out a number of sources, only some of which will be appropriate to any one system.

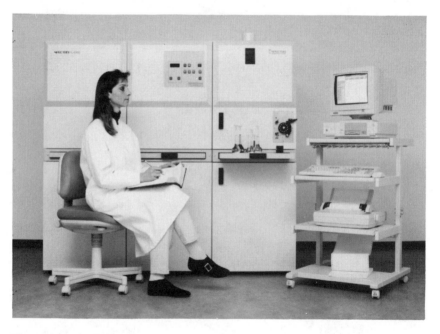

Figure 7.41 An ICP spectrometer (Spectro).

There are two very important points to note here. Depending on the type of spectrometer used

1. there may be a limit to the size of particle which can be detected and
2. it may be possible to determine the 'compound' rather than just the elements present.

Table 7.13 Elemental sources

Element	Possible source
Aluminium	Atmosphere, bearings, pistons
Antimony	Bearings, grease
Barium	Additives
Boron	Additives, cooling water
(Brass)	(See copper, tin and zinc)
Calcium	Additives, quarry, sea water
Chromium	Cylinders, piston rings, seals
Copper	Bearings, bushes, coolants, cylinder liners, piston rings
Iron	Bearings, cylinders, gears
Lead	Bearings, fuel, grease, paints, seals, solder
Magnesium	Additives, shafts, valves
Molybdenum	Additives, piston rings
Nickel	Bearings, gears, turbine blades, valves
Phosphorus	Additives
Silicon	Additives, atmosphere, gaskets, grease
Silver	Bearings, shafts
Sodium	Additives, coolants, grease
Tin	Additives, journal bearings, seals, solder, worm gears
Titanium	Bearings, paint, turbine blades and discs
Vanadium	Fuel, oil, catalytic fines
Zinc	Additives, bearings, coolant, grease

Thus atomic absorption and atomic emission (including the inductively coupled plasma) are normally restricted to particles of maximum dimension 10 µm (or even much less with certain models) – and larger particles are just not detected at all. X-ray fluorescence (XRF) is non-destructive and hence can 'see' any size particle. Energy-dispersive X-ray (EDX) analysis can be highly selective in undertaking spot checks on individual particles, again non-destructively. FTIR instruments (Fig. 7.42) are primarily for oil analysis *in toto* and do not look at the elements so much as at the spectra associated with an oil when new compared with those obtained when it is contaminated. Mass spectrometry also looks at the compounds, i.e. the oil, rather than at the particles which may cause a problem.

Table 7.13 shows how the analysis can identify problems associated with the oil condition, the ingested contaminant and the debris content, depending on the elements detected. Historically, spectrometric oil analysis (SOA) has been seen as a major wear debris technique. Although it is so valuable, it has been misused quite frequently through a lack of the understanding of its limitations (as mentioned in the previous paragraph).

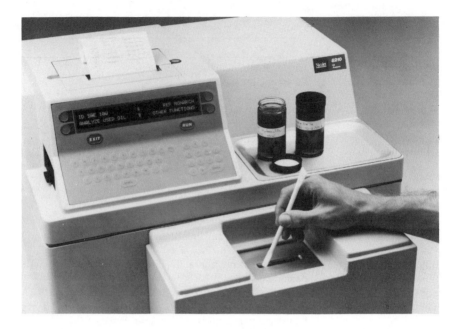

Figure 7.42 An FTIR oil analyser (Nicolet).

7.6 THERMOGRAPHY

7.6.1 Introduction

Temperature is a key condition indicator. It applies to the human body and it applies to machines. Both a rise in temperature and a fall in temperature can be critical. In other words, it is important that a correct working temperature is present, and if it is off limits then it is likely that something has gone wrong.

One use of temperature detection is that of thermodynamic efficiency, where, by sensing the rise in temperature of fluid as it passes through a pump, the efficiency can be monitored. This is described in more detail in Chapter 8.

The single rise in temperature of a product, such as an engine block, can again be an important monitor. This is mentioned in section 7.9. However, such temperature changes cannot be seen with the naked eye – unless they reach 750°C!

Thermography is a much more comprehensive and detailed monitoring method and, because of the electronics used, it is able to 'see' surface temperatures as low as −40°C, or even lower.

7.6.2 Brief description (Fig. 7.43)

'Thermography' is basically the remote measurement of the temperature distribution of an object in real time, making it optically visible in an image of different colours (or grey scales). It is the detection of the temperature pattern of a machine or mechanism while it is in operation. In simple terminology, it could be considered as 'looking for the hot spots' as they occur. More explicitly, it is 'observing the subtle changes in temperature generation and transfer' as a machine changes its behavioural characteristics.

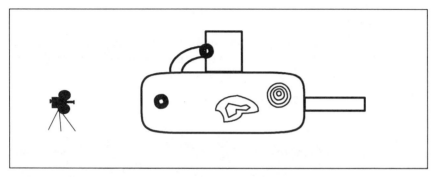

Figure 7.43 Thermography.

7.6.3 Practical applications

The practical applications are anywhere where temperature variations have happened because of a fault developing. The initial temperature variation may be due to the fit of two components, to the insulation between them, to friction, etc. There are many and various examples where these features have changed with wear or deterioration. The following are a few of those known to the author:

- loose connections on electrical wiring;
- corroded joints on machine parts in the marine industry;
- uneven wear in refractory lining;
- blockage of furnace tubes;
- hot spots in furnace reformer tubes;
- switch gear collapsing;
- high voltage transmission line connectors deteriorating;

- poor combustion on incinerator burners;
- excessive wear or fatigue on bearings.

In essence, where there is any mechanical, thermodynamic or electro-magnetic process which results in local heat generation, thermography can detect changes due to machine condition deteriorating.

7.6.4 Detailed discussion of thermographic measurements

In thermography, the invisible radiation spectrum may be converted into a grey scale or to a visible colour spectrum, where each colour represents a different stated intensity of radiation (temperature). An example of a grey scale image is shown in Fig. 7.44 (although this would be much clearer in colour). The range of temperature on the motor spindle is normally from 36°C to 45°C so the overheating of the bearing at around 70°C is easily visible.

The colours used to describe temperatures may not always follow the usual optical spectrum in order. Two examples, A and B, in Table 7.14 show the change in colour with decreasing temperature.

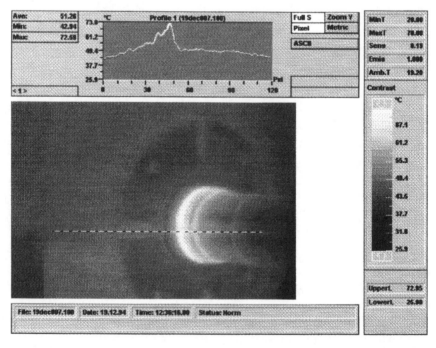

Figure 7.44 Thermal image of an overheating bearing on a shaft spindle (GoRaTec).

Table 7.14 Examples of different changes in colour with decreasing temperature

A			B
Red	Strong thermal radiation	Higher temperature	White
Orange			Yellow
Yellow	Average thermal radiation	Average temperature	Orange
Green			Red
Blue	Low thermal radiation	Lower temperature	Purple
Indigo			Blue
Violet			Indigo

The standard technique is that of infra-red detection incorporated within a portable camera with a video recorder to store the images. In some cases a second camera system is included in the same module; this means that both the thermal image and an optical view of the machine can be imaged at the same time (this can be helpful in identifying which component is suspect).

The camera is capable, with its multicolour range, of detecting an enormous range of temperature with considerable detail. This detail provides the necessary clarity to a thermographic image. (Some cameras have less definition because of a reduced range of colours.) One of the convenient ways of using the camera system is shown in Fig. 7.45, in this case like a portable video camera.

In 1994 typical temperature rises in a working system involving electrical installation were given by Agema for electrical installations as follows:

- minor problem, 1°C–10°C;
- problem, 10°C–35°C;
- serious problem, 35°C–75°C;
- critical problem >75°C.

The maintenance of the absolute temperatures (and hence colours portrayed) depends on the emissivity, ambient temperature and object distance, all of which can be compensated for in the more advanced systems.

In extremely delicate situations the energy emitted by the normal thermographic camera may affect the surface under measurement and distort the temperature pattern; in this case, a low-energy technique must be used.

- One low-energy method is pulsed thermography which uses flash-lamps to apply short sharp bursts of energy; in this case it may be necessary to eliminate carefully different emissivities across the component's surface.

Figure 7.45 Remote thermography in practice (Agema).

- Another method uses a harmonically modulated energy source; here there is a feedback control to produce a pure sine wave synchronized with the scanner frequency, eliminating the need for emissivity checks; it is also independent of reflected ambient radiation. Localized dark or light spots in the resulting on-screen images indicate changes in the surface being examined.
- A further method, where very fine detail is being examined, is to use a microscope where the microscope lens magnifies the object optically up to a resolution of $25\,\mu m$, enabling the local temperature to be measured. In other words the microscope has isolated the energy from the camera.

Another problem can occur if there are strong electromagnetic fields present, such as in copper refineries or aluminium works. In this case the normal cathode ray tube (CRT) viewfinder cannot work successfully and an alternative, such as the colour LCD viewfinder, must be used. Incidentally the use of a colour viewfinder means that all items of interest remain within the temperature range and do not go into image saturation (a condition which is harder to spot in grey scale images).

Table 7.15 Possible specifications of thermographic equipment

	Specification	*Examples*
What can be done	Range of temperature	0°C to 800°C, –40°C to 800°C, to 2000°C
	Thermal resolution	0.05°C, 0.15°C, 0.5°C at 25°C target
	Accuracy	Full scale ± 0.4%, ±2% reading
	Resolution	12 bit (4096 levels) or 8 bit (256 levels)
	Pixels per line	1.5 mrad, 2.2 mrad (0.125°)
	Lines per image (vertical IR line)	300, 320 (horizontal) × 240 (vertical) 200
	Live and freeze modes	Both
Ease of use	Colour viewfinder	½ in LCD viewfinder
	Image set-up	Automatic
	Lens focus range	200 mm – ∞
	Field of view (FOV)	15° × 10° (see Fig. 7.46)
	Zoom facility	Yes (2×), (6×), no
	Portability	3.8kg, 248×113×241mm
How it works	IR detector	HgCdTe, InSb (10 elements)
	Operating wavelength	3–5µm, 8–14µm
	Lens aperture angle	2.5°, 5°, 7°, 10°, 12°, 15°, 20°, 40°
	Lens focal length	25mm, 50mm, 75mm
	IR field rate (scan speed)	15, 20, 30Hz (frames s⁻¹)
	Frame frequency	0.25 to 0.8s
	IR cooling	Ar gas cooling, none
	Compensation	For distance, ambient temperature and emissivity
How the results are presented	Range of grey scale palettes	3 and reverse
	Range of colour palettes	3, 5
	Output	GPIB, RS-232C
	Monitor	4in LCD monitor
	Visible image available at same time	Integral, on separate camera, superimposed
	Storage of image	3½ in floppy disc, CD
	How many images per disc	40, 70
	Spot temperature read out	Yes

7.6.5 How to start

A thermal imaging system having been purchased, or hired, it needs to be set for the temperature range required; then the start button can be pressed.

In fact there may be a little more complication, namely in deciding which model is best. For instance the specifications given in Table 7.15 present a considerable number of opportunities.

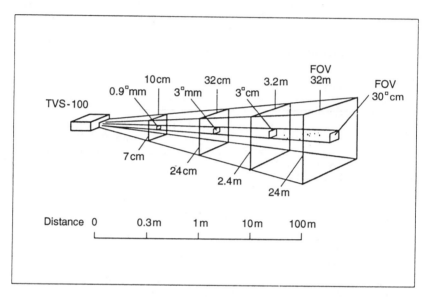

Figure 7.46 Field of view for 18.4° horizontal, 13.5° vertical angles (GoRaTec).

To choose camera and recorder specifications is not easy. It depends on what it is hoped the camera will find. Undoubtedly, for very detailed machinery, the most valuable asset will be a clear resolution. On the other hand, for examination of more general regions, the angle of view may be the most important (Fig. 7.46). One must also consider how close one can get to the machine or component; the high range electro-optical zoom on the Jenoptik camera system can cover 6:1 without changing objectives.

7.7 LEAKAGE

Leakage, in this context, involves the loss of something which is volatile or fluid. For instance, the following are all constituents of leakage, but in very different forms (Fig. 7.47):

- liquids;
- gases;
- electrical currents;
- information.

While leakage of current may be seen as an electrical problem, it does affect mechanical systems and should therefore be considered here. Each of the first three, liquid, gaseous and current, has a different measurement concept and, hence, is dealt with separately in the next few pages. 'Information leaks' are rather different on two counts – they may well be beneficial (as would be the contents of this book, if leaked) and, secondly, they are not mentioned further.

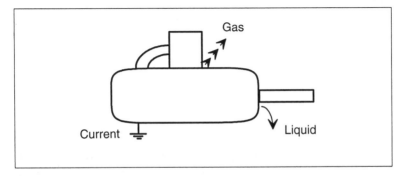

Figure 7.47 Leakage from a machine.

7.7.1 Liquid leakage

Liquid leakage may be unavoidable in a system where lubrication is essential, such as from the seals of an actuator or from other dynamic seals. However, of course, sufficient leakage for lubrication is not the same as wasteful leakage; hence the rate of leakage is important. Leakage may be within the machinery and hence not detectable on the outside; this would be, say, where a high pressure pump loses output back to the input owing to leakage across the port plate. Note that liquid leakage can occur in reverse, namely water could get into the working oil, or oil could get into the coolant water.

7.7.2 Brief description of liquid leakage

Liquid leakage occurs where liquid is permanently lost from a machine, either by dripping or by flowing out of, or into, certain parts of the machine. It is considered serious where the flow is such as to be greater than that required for lubrication purposes, or where it changes the characteristics of the machine function.

7.7.3 Practical applications of liquid leakage

Liquid leakage occurs at

- pipe joints,
- welds,
- adapters,
- reservoirs and tanks,
- motors,
- valves,
- hoses,
- couplings,
- cylinders,
- caps and plugs,
- pipes,
- hose fittings,
- seals,
- pumps and
- components which have to be replaced, such as filters.

In the last-mentioned case, the leakage may not be so much as from the component but rather from the process of removal and refitting. Strip and rebuild is always prone to cause disturbance to threads and connectors.

7.7.4 Detailed discussion of liquid leakage measurements

Before discussing the detailed types of leakages and how each is detected and analysed, it is worth considering the various approaches which can be taken. It is the human analogy which can help here – how do we, independent of any other monitor, detect liquid leakage?

- hear – particularly liquid hissing through a small orifice or crack;
- touch – liquids are wet to the touch;
- sight – particularly gross liquid leakage;
- smell – particularly liquids which have become contaminated;
- taste – particularly apparent on a finished consumable product.

We will find that most leakage techniques are based on our five senses, in some cases with considerable lateral thought. The following relate to water losses:

- noise of a leak – hear;
- ground water logged – touch;
- subsidence or extra grass growing – sight;
- biological activity in static water – smell;
- contamination into the water – taste.

In addition we could include two other features which are commonly monitored in water systems:

- unusually high consumption – performance;
- loss of pressure – performance.

Liquid leakage may occur as a result of poor manufacture (out of specification or tolerance), poor fitting (as mentioned earlier), incorrect seals, overload or overpressure, vibration or some other damaging influence on the machinery. When a complex component is built, and it involves a pressurized system, it is common to insert it in a wet bath, to pressurize it, and to see whether any bubbles are emitted. That covers the initial build of the product, but, as regards condition monitoring of the machinery in use, there are a variety of ways of detecting leakage. Table 7.16 lays out some possibilities, including those mentioned earlier.

The British Standard BS 7388: 1991 (*Guide for prevention of leaks from hydraulic fluid power systems*) gives some guidance categories for the importance of leaks from a system, in three levels of importance (Table 7.17).

Pressurized liquid systems may be examined for leaks by using air or a suitable tracer gas (i.e. air seeded with a 'scent') as shown in Table 7.18 (derived from Ion Sciences Limited, Cambridge).

Another technique is to use an ultrasonic exciter on the inside of a container under test and to place a detector on the outside to see what and where the ultrasonic waves escape.

7.7.5 How to start with liquid leakage

Liquid leakages are very common. The main problem, however, is anticipating where they may occur. Initially a visual observation may be arranged on a regular basis. Once some experience has been built up, then a more permanent sensor can be fitted.

7.7.6. Gaseous leakage

Gaseous leakage may not necessarily be of great loss in itself, but it may show a deteriorating situation developing. It can also have serious

Table 7.16 Detection of liquid leakage in operating systems

Type of detection	Comments
Visual observation	Frequent visual checks are necessary on a regular basis if this is to be acceptable
Fluid loss assessment on drip trays or in collection tanks	This is visual observation of the leakage where it is expected to occur
Amplified visual observation	If leakage is expected to occur, the container or pipe carrying the liquid can be coated with an 'optical brightener'; any liquid coming out will then be more obvious due to fluorescence (this is particularly valuable in underwater applications)
	An extension of this for ground based containers is the adding of a small amount of chemical tracer to the liquid; when the liquid escapes, the surrounding land begins to give off a hydrocarbon vapour which is detectable
	A further application is the detection of non-synthetic oils which leak into water – they already have fluorescent characteristics and hence can be detected by a fluorescence meter
Fluid level change in reservoirs	Note that not all changes are caused by leakage; fluctuations may occur owing to system behaviour, e.g. cylinder action and temperature changes
Fluid weep detectors	By using a suitably placed microchip which is sensitive to moisture, or a wrapping of moisture sensitive tape, or the shorting between two continuous sensitive wires in a cable, the leak can be sensed immediately
Fluid mist detectors	Only suitable in an enclosed space, but particularly successful where small hole leakage may occur in a high pressure system
Loss in system pressure or flow	Pressure sensors or flowmeters may need to be placed in various positions in the system to determine at what point leakage is occurring (Fig. 7.48)
Change in fluid constituency	Oil-in-water or water-in-oil monitors can be used; these are discussed in section 7.5 (Table 7.12)
Listening to leak noise	Audio frequency listening may be possible for the larger leaks and the occasional smaller leaks. A simple listening prod can be used (like the old engineer's screw driver); however, high- and low-pass filters will help isolate the leak noise from other noise in the vicinity (e.g. 340 Hz to 625 Hz). A series of digital filters can be even more discriminating (e.g. from 20 Hz to 2.5 kHz)

Table 7.16 (contd)

	Another way to determine the position of a leak is to mount two sensors along a line, either side of the suspected leak; correlation of the two signals will then indicate the proportional position of the leak (Fig. 7.19) (Note that three sensors may be necessary if the position is obscure)
Ultrasonic listening	This is a major use of ultrasonic detectors (section 7.3); it is primarily used where the system is under pressure, and a leak is a 'spurt' rather than a 'drip'

Table 7.17 Levels of importance of liquid leakage

Leakage source		Category		
		1	2	3
Rotating shaft seal	In motion	No drips	Maximum: 1 drip h^{-1}	Maximum: 1 drip $(10\,min)^{-1}$
	Stationary	No drips	No drips	Maximum: 1 drip h^{-1}
Reciprocating shaft seal	In motion	No drips	Maximum: 1 drip $(30\,min)^{-1}$	Maximum 1 drip $(5\,min)^{-1}$
	Stationary	No drips	No drips	Maximum: 1 drip h^{-1}
Pipe connection		No drips	No drips	No drips
Other static seals		No drips	No drips	No drips
Swivel joint	In motion	No drips	Maximum: 1 drip $(30\,min)^{-1}$	Maximum: 1 drip $(5\,min)^{-1}$
	Stationary	No drips	No drips	Maximum: 1 drip h^{-1}

Table 7.18 Leak assessment guidelines (by gas)

Product	Test pressure (bar)	Leak	Test method
Water cooling systems	1–2	4–7 mL min^{-1}	Air pressure decay
Oil pipelines etc	0.1–7	6–15 mL min^{-1}	Air pressure decay or flow decay
Refrigeration and air conditioning plant	2–20	5–15 g $year^{-1}$	Tracer gas
Electrical housings and connectors	0.1–1	0.01–1 ml min^{-1}	Air pressure decay or tracer gas

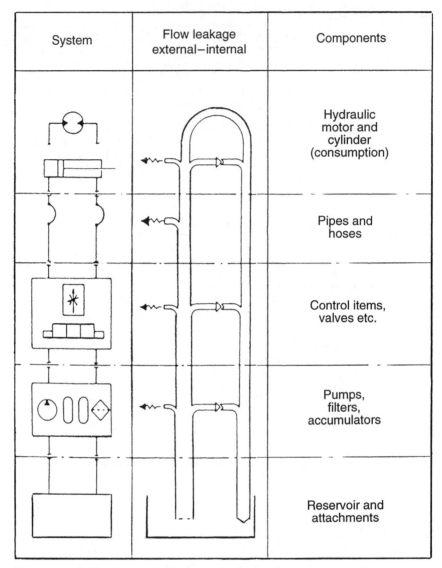

System	Flow leakage external-internal	Components
		Hydraulic motor and cylinder (consumption)
		Pipes and hoses
		Control items, valves etc.
		Pumps, filters, accumulators
		Reservoir and attachments

Figure 7.48 Determination of leakage in a hydraulic system with flowmeters.

health and hazard consequence. As with liquid leakage, it can occur in the opposite direction, such as with air getting into the system or process fluid.

7.7.7 Brief description of gaseous leakage

Gaseous leakage occurs when air, or a process gas, undesirably exhausts to atmosphere or into the wrong cavity.

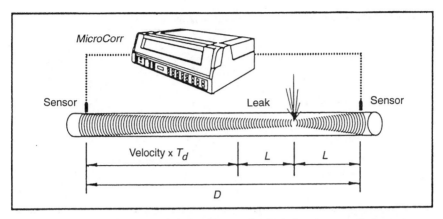

Figure 7.49 Determination of the position of a leak by the correlation of two 'listeners' (MicroCorr – Palmer Environmental Services). The leak position L is given by $L = D - VTd/2$, where D is the distance between sensors, V is the velocity of sound for the pipe under test and Td is the time delay (transit time difference).

7.7.8 Practical applications of gaseous leakage

Gaseous leakage may occur from any system which involves pressurized gas pipes or pressure vessels.

7.7.9 Detailed discussion of gaseous leakage measurement

Again, it is helpful to think of the human analogy – how do we, independent of any other monitor, detect gaseous leakage?

- hear – particularly gas issuing through a small orifice or crack;
- touch – is not relevant to gaseous leakage;
- sight – as a secondary feature such as the movement of a flapping label;
- smell – particularly true of leaking gas;
- taste – particularly a finished consumable product.

Gaseous leakage may occur from fractures of any pipe line or pressure vessel, mainly due to fatigue or overload. Misuse, in the sense of inadequate tightening, may also be a contributing factor.

Gaseous leakage may be detected in a small number of ways as shown in Table 7.19.

Gas leakage techniques may also be used for the examination of pressure systems which may eventually be used for liquids – see the sections above on liquid leakage.

Table 7.19 Types of detection of gaseous leakage

Type of detection	Comments
Visual observation with fluid	A soapy solution applied in the vicinity will bubble or froth as the gas passes through it
Loss of system pressure or flow	A drop in pressure or flow indicates a leak. Automatic automobile tyre pressure sensing is now undertaken in some coaches, so that the driver knows whether the tyre is becoming dangerous (Fig. 5.3)
Listening to leak noise	See comments in Table 7.16
Sniffing the gas	Although some gas detectors are general purpose (for all gases), accurate detection will be limited to the specific sensors used, e.g. hydrogen, helium, methane, carbon dioxide, argon, oxygen etc. (Table 7.3)

7.7.10 how to start with gaseous leakage

Consider what gases are involved and possible points of entry or exit and choose an appropriate monitor suitable for the application.

7.7.11 Current leakage

Current leakage is a loss, because it has to be paid for. It is something which is certainly undesirable and may lead to further machine trouble.

7.7.12 Brief description of current leakage measurement

In simple terms, current leakage occurs when insulation breaks down and there is a flow of current through an unintended conductor to earth. However, in modern electronic systems, the earth acts as a key working component of the electrical installation and therefore some leakage is expected, and indeed intended.

7.7.13 Practical applications of current leakage

Excess current leakage (at 50 or 60 Hz) occurs from faulty or damaged electrical supply lines. One example is the traction industry or, indeed, from any machinery where power lines could be damaged.

Additional leakage at harmonics of the mains 50 or 60 Hz frequency occurs when the AC power supply to electronic equipment is 'filtered' to remove disturbances. Where switch mode technology is used, the leakage will appear at very high frequencies (up to 100 kHz) and will not be synchronized to the mains frequency. The maximum acceptable (permissible) earth leakage for portable PCs is stated as being 3.5 mA.

7.7.14 Detailed discussion of current leakage measurement

While not really the subject of this book it is worth mentioning that direct current imbalances may be detected using a Hall effect current transformer or, where there is temperature instability, by means of a differential current transformer. (This is also discussed in section 6.2.) The design of such a sensor might have to cover a range of specifications such as the following:

- current, 0–100 A at both 50 Hz and 60 Hz to cover single- and three-phase motors;
- response time better than 0.2 ms (DC) and 25 ms (AC);
- linearity, 1% of full scale over the full operating environmental temperature range.

It must be remembered that such monitoring not only senses the machinery condition but also checks on the safety of the equipment. The use of steel wire armoured (SWA) cable is common for an earth path; this is usually totally acceptable for mains frequencies but it is not suitable for the kilohertz frequencies. At the high frequencies, steel wire armouring will have significant impedance to the current and voltage differences will develop with resulting electric shocks and electronic output faults.

If an earth fault develops, and extension leads are in use, it is possible for exposed metalwork to go live at a potentially lethal voltage. If working equipment is earthed via the data leads of peripheral equipment, and then the working equipment is unplugged, the earth pin on the plug becomes a serious risk to human life.

7.7.15 How to start with current leakage

Current leakage must be avoided as much as possible, but, if there are susceptible regions, then a suitable portable clamp-on or permanent current monitor can be fitted.

7.8 CORROSION

7.8.1 Introduction

Not only is corrosion the curse of the static edifices built by civil engineers, it is also a major headache in machinery. It has been reported that 50% of forced shutdowns in steam generating plants in the USA (costing $3 billion per year) are caused by corrosion. It is considered that between 3% and 5% of GNP world-wide is lost in this way. Anti-corrosion additives in oils and fluid power water are now considered essential.

One problem with corrosion is that it is not uniform. For instance, we all know that rust is caused by the mixture of water and air, and also that the corrosion is accelerated if there is a salty atmosphere such as near the coast or when roads have been prepared for impending snow or ice. These are not the only factors; there is temperature and pressure, there are certain catalysts and then the acidity or alkalinity of the process in hand. Air temperature, for instance, has an effect in moving the dew-point, and hence the local temperature at which condensation takes place (Fig. A.4).

Rather surprisingly, bacteria also play a part. To estimate corrosion is therefore extremely difficult, and to be able to monitor it is essential and will certainly bring considerable financial benefits.

Corrosion occurs because of

- a corrosive atmosphere around an unprotected component,
- electrochemical corrosion due to two dissimilar reactive metals in contact,
- a chemical reaction due to combustion,
- bacteria which break down nitrites (nitrites are used in anti-corrosion inhibitors) and
- bacteria which break down sulphates (and which produce the acidic hydrogen sulphide).

These bacteria types impose an interesting problem. They are not harmful to humans but are harmful to machinery, particularly pipework carrying water and the very additives put in to stop corrosion provide the 'food' for the bacteria to grow and do their deadly work.

Erosion, a complementary action to corrosion, occurs because of

- cavitation,
- fatigue stress and
- temperature fluctuations (fatigue).

In some ways the different types of corrosion and erosion are similar; and certainly there is an overlap between them. For example, cavitation can be described as a fatigue effect where very small vapour bubbles on the surface of a component collapse with chemical or electrochemical surface activity. It is usually due to insufficient pressure at the surface and can occur in marine or fluid power hydraulic systems (section 7.4.3).

7.8.2 Brief description

Corrosion is unwanted loss of material through chemical or electrochemical action (Fig. 7.50). A complementary subject is erosion through fluctuations in pressure, stress or temperature. Indeed, erosion is sometimes termed 'flow-assisted corrosion'.

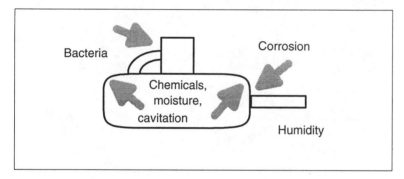

Figure 7.50 Corrosion in machinery.

7.8.3 Practical applications

Corrosion can be measured over an enormous range of rates. For instance, measurement down to a scale of 0.2 mm year^{-1} is not uncommon. On the other hand, considerably higher rates are also involved. Monitoring in corrosive, or semicorrosive, atmospheres is one example such as with outlet ducts and stacks where the flue gas desulphurization system is used in power generation plant. Nuclear, oil and gas, chemical and petrochemical industries are all prone to such corrosion. Other industries include aerospace and subsea.

Most fuels provide a certain amount of corrosion. Sulphuric acid mist can be produced at 150°C in gases, condensing on cold surfaces. Conversely, a very low sulphur content in the fuel itself (in marine diesel engines) can cause high liner and piston ring wear owing to an inability to neutralize the highly alkaline lubricants.

Erosion by cavitation could affect propellers, rudders, pumps, motors as well as plain bearings and diesel engine cylinder liners (coolant side).

Figure 7.51 shows the kidney port face of the manifold of a high pressure pump which has been run at too low an inlet pressure. Cavitation erosion has occurred radially, rather than in the usual circumferential direction, owing to the case pressure being low. This erosion caused a considerable drop in efficiency.

Figure 7.51 A pump pressure face showing cavitation erosion.

7.8.4 Detailed discussion of corrosion measurements

Corrosion and erosion detection may be undertaken along the lines of

- observation of cracks,
- determination of how much metal has been lost,
- assessment of the likelihood of chemical reaction (electrochemical techniques) and
- assessment of the presence of nitrite-reducing, or sulphate-reducing, anaerobic bacteria.

Corrosion monitoring starts with a determination of how much material has been lost, but of much greater value is the determination of the rate of material dissipation at the current time – i.e. on-line monitoring in real time. There is a great deal of discussion about the relative merits of the different techniques; the following list, and the brief descriptions later (with acknowledgement to a number of companies and in particular, Dr Patrick Stokes of Real Time Corrosion Management Ltd), should give sufficient background information:

- crack monitoring by ultrasonics, eddy current, magnetic particle inspection (MPI), magnetic flux leakage or liquid penetrant;
- the weight loss method (with a coupon);
- thickness measurement (with or without a coupon);
- electrical resistance (ER);
- thin layer activation (TLA);
- zero resistance ammetry (ZRA, or galvanic coupling);
- linear polarization, or DC polarization (linear polarization resistance measurements – LPRMs);
- electrochemical impedance monitoring (EIM);
- electrochemical impedance spectroscopy (EIS);
- electrochemical noise (EN);
- harmonic analysis (HA);
- field signature method (FSM);
- biological activity monitoring.

The British Standard BS 6918:1990 and International Standard ISO 8044:1989 (Glossary of Terms for Corrosion of Metals and Alloys) give a number of useful definitions of corrosion. Some of these are reproduced in Table 7.20. Some corrosion probes are shown in Fig. 7.52.

Table 7.20 Types of corrosion (BS 6918:1990)

Corrosion type	Description
Atmospheric corrosion	Corrosion with the earth's atmosphere at ambient temperature as the corrosion environment
Corrosion fatigue	A process involving conjoint corrosion and alternating straining of the metal. Note that corrosion fatigue may occur when a metal is subjected to cyclic straining in a corrosive environment. Corrosion fatigue may lead to cracking
Electrochemical corrosion	Corrosion involving at least one electrode reaction
Fretting corrosion	A process involving conjoint corrosion and oscillatory slip between two surfaces in contact. Note that fretting corrosion may occur, for example, at mechanical joints in vibrating structures
Gaseous corrosion	Corrosion with gas as the only corrosive agent and without any aqueous phase on the surface of the metal
Polarization resistance	The quotient of electrode potential increment and current increment. Note that usually the polarization resistance is measured in the vicinity of the free corrosion potential (i.e. the electrode potential of a metal in a given corrosion system)

Figure 7.52 A selection of corrosion probes (Real Time Corrosion Management).

(a) Crack monitoring by ultrasonics, eddy current, magnetic particle inspection, magnetic flux leakage or liquid penetrant

This set refers to the classic non-destructive testing (NDT) of components. Each technique is basically a means of detecting a crack which has already developed. They do not normally show the rate of corrosion, and would be considered only as a simple means of detection if nothing else is available. They are described in more detail in NDT manuals.

(b) The weight loss method (with a coupon)

A sample of the material in use (a 'coupon' in the form of a strip or disc) is carefully weighed and put in the situation where corrosion is expected. (This may be in a simulated situation but the real environment is preferable.) After a convenient period of time, say 3 months, the coupon is removed and re-weighed. Any loss in weight will convey the rate of material loss over the time since the previous weighing. (Additional notice is taken of the surface appearance and presence of deposits on the coupon.)

This is known as an 'historical' method, i.e. it shows what has happened in the past but does not reveal what the current or intermittent situation is. It is retrospective and cumulative, and can be used (in spool piece designs) for inserting into side streams for the assessment of pitting and stress corrosion, etc.

Note that the coupon should normally be electrically isolated from the plant, in order to prevent galvanic corrosion between the coupon and plant.

(c) Thickness measurement (with or without a coupon)

Complex geometries such as bends in pipes, valves or nozzles can be manually scanned to produce accurate material thickness related images. This can be done using ultrasonic transducers.

(d) Electrical resistance

ER probes use a bridge circuit to measure the change in cross-sectional areas resulting from corrosion. Designs are available as wire, tubular or flush mounted.

This technique normally only reveals what has already happened, and is used for long-term monitoring on site where a gradual trend can be detected. It is best in carbon-manganese steel systems.

(e) Thin layer activation

Although expensive to apply – a specific region normally has to be irradiated at an atomic energy establishment – the detection of a change in γ radiation emitted (on the outside of a pipe, say) is exceptionally accurate. It is continuous and indicates the cumulative situation. It is good for carbon steels.

(f) Zero resistance ammetry (or galvanic coupling)

ZRA looks mainly at the galvanic coupling between dissimilar metals, e.g. copper and steel, and monitors the current flow between them. This changes with corrosion and is particularly useful in those applications involving a variation in oxygen contents in the liquid, e.g. water. However, it is less sensitive at low corrosion rates.

(g) Linear polarization, or DC polarization (linear polarization resistance measurements)

This is the basic type of electrochemical technique; it uses an inserted probe. It can indicate the likely rate of corrosion and the pitting tendency of a component in a corrosive fluid (which is conducting). Note that the electrode within the probe must be of similar material to the material used in the real situation.

The slope of the line of electrode potential versus test current (in the region of the corrosion potential) enables a corrosion rate to be determined.

It is particularly useful for carbon steels and copper alloys where the corrosion rate is uniform, or where only a general corrosion rate is required. Care must be taken to ensure that the system has reached a steady state (or is suitably corrected), otherwise overestimates of corrosion rate will be achieved. On the other hand, scales or deposits cause an increase in resistance and hence an underestimate of corrosion rate.

(h) Electrochemical impedance monitoring

This is an advance on the LPRM. Although it is limited to an overall reading, like the LPRMs, it can be used in lower conductivity electrolytes than LPRM. It examines the electrochemical response for a limited frequency distribution. The general determination of corrosion by EIM is valuable in assessing corrosion rate differences with process changes; a total life prediction can also be made – very suitable for machine monitoring. Some indication of the corrosion mechanism may be revealed.

(i) Electrochemical impedance spectroscopy

While the addition of a small AC signal applied to the corrosion probe in an otherwise DC polarization device will bring a greater sensitivity as the electrical impedance is measured, an even greater improvement can be made if the frequency of the AC signal is varied over a much wider range, such as 20 kHz to 1 MHz. This is EIS.

EIS is able to distinguish between different types of corrosion. The probes can be installed directly into the system as flush-mounted multi-electrode probes to enable continuous real-time monitoring to measure the actual corrosion during operation. In itself, the test technique is non-destructive and can show the opposite, namely the build-up of barrier films.

The description of HA in section 7.8.4(k) should also be referred to.

(j) Electrochemical noise

This is the measurement of the low frequency, low amplitude, random fluctuation of corrosion current and corrosion potential produced during a corrosion process. The type of fluctuation, or transient, characterizes the form of localized corrosion, while correlation of the two signals provides an estimate of the corrosion rate.

This technique is able to cover a high range of conductivity; for instance, it is suitable for both salt water systems (with high conductivity) and the much lower conductivity regimes where other techniques cannot manage. It is particularly sensitive to the onset and propagation of localized attack which may be microbial. A comparison is undertaken where the corrosion rates are examined both before and after biocide treatment is applied to the fluid. The instrumentation is often able to detect the effect of oxygen content changes.

This can be combined with EIS.

(k) Harmonic analysis

HA is an extension of impedance monitoring. Because there is not really a linear relationship, but rather a slightly curving one, between electrode potential and test current (as discussed for LPRM in section 7.8.4(g)), that when a sinusoidal voltage is applied harmonics will be generated. A simple corrosion rate can be determined from a calculation involving the first three harmonics.

(l) Field signature method

All steel structures have their own unique 'fingerprint' when an electrical current is applied. With FSM a series of pins are attached to the outside of the metal component in the form of an array over the area to be monitored. An electrical field pattern is then derived for the 'new' component. As the metal begins to corrode so the pattern is modified as seen by graphical plots; these plots can determine not only the severity and location of the corrosion, but also the trends and rate of metal loss.

(m) Biological activity monitoring

The presence of anaerobic bacteria can be detected using culture tubes; these are known as 'Sig' tests. Separate tubes are available for the nitrite-reducing bacteria and for the sulphate-reducing bacteria. Severe infection should be indicated by a blackening in the tube after approximately 24 h; slight infection will be shown after 48 h.

An alternative way is to use electrochemical noise, as described in section 7.8.4(j).

7.8.5 How to start

Before choosing a monitor, decide the type of corrosion or erosion which might possibly take place. A selection of monitors should be tested in order to determine the optimum.

(a) Corrosive atmospheres

Are any corrosive mists likely to be present?

(b) Electrochemical combinations

Consider what combinations of potentially damaging elements and compositions are present.

(c) Combustion reactions

Which parts of the system are involved in combustion and hence will expect high temperatures mixed with corrosive gases from time to time?

(d) Anaerobic bacteria corrosion

('Anaerobic' means that the bacteria do not need oxygen for their growth.)

Is sodium nitrite or borax present in the water (as a corrosion inhibitor)? If so such bacteria as the nitrite-reducing bacteria would be provided with an encouragement to get to work, their numbers would increase and there would be a greater demand for increased levels of the inhibitor.

On the other hand oil and gas industries would expect to see high levels of sulphates, and there the sulphate-reducing bacteria would thrive.

(e) Cavitation erosion

Are there places in the system where negative or low pressures may occur – such as in the approach to a pump inlet?

(f) Fatigue and stress erosion

Where in the system will there be fluctuating stresses?

7.9 STEADY STATE ANALYSIS

7.9.1 Introduction

The term 'steady state' analysis originated in comparison with the two other prime condition monitoring techniques – dynamic state (i.e. vibration, ultrasonics etc.) and fluid state (i.e. spectrometric, oil properties, wear debris etc.). Steady state covered all the slowly varying parameters which could be seen conventionally on dial gauges, such as pressure and temperature. In some ways such monitoring is similar to performance (which is covered in the next chapter); however, it is different in that it does not require a total assessment of the machine efficiency but rather just an individual detection of usually just one feature which can convey a deterioration.

Steady state features of machinery could include the following:

- colour (section 7.1);
- corrosion (section 7.8);
- crack (section 7.8);
- current (section 6.2);
- density (section 7.5);
- displacement (section 7.9);
- efficiency (section 8.1);
- energy consumption (section 8.1);
- fatigue (section 7.2);
- flow (section 7.9);
- leakage (section 7.7);
- level (section 7.9);
- load;
- performance (section 8.1);
- pressure (section 7.9);
- rotation (section 7.9);
- smell (section 7.1);
- speed (section 7.9);
- stress;
- temperature (section 7.9);
- thickness (section 7.8);
- torque (section 6.2);
- weight.

This section deals with just five of these features, but, from these five, the reader should be able to sense how to approach the subject for any of the others appropriate for his or her application. (Further details on others are given in the sections quoted.) The five features covered here are

1. pressure,
2. flow,
3. level,
4. speed (including rotary displacement and rotation) and
5. temperature.

7.9.2 Brief description

A steady state parameter is monitored to detect significant changes which may indicate a deterioration in the condition of the machine (Fig. 7.53).

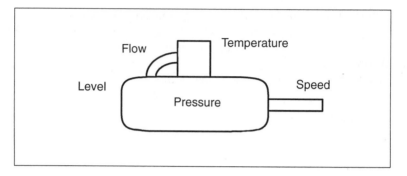

Figure 7.53 Steady state monitoring.

7.9.3 Practical applications

Almost all machine applications have a relevance here. Although the use of a 'dial gauge' has been mentioned above, digital values may be produced by a transducer and hence more distant sensing is possible which can be fed directly to some logging device or to a remote station (Chapter 9).

It must be remembered that there are many different types of monitors. Some are very suitable for one type of machine application, and others are better for another. However, the advantage of the steady state monitors is their low cost, particularly for machine systems already in place. Pressure gauges, for instance, may already be operational, and all that is needed is that extra awareness of what those gauges reveal either from spot checks or from long-term logging of their values.

Typical systems which can see positive results from this monitoring would be

- process plants,
- pressurized systems,
- lubricated systems and
- drive systems.

7.9.4 Detailed discussion

Each of the five steady state features will be described in detail, with a range of possible techniques or devices which are available.

(a) Pressure

Pressure may drop for many reasons, such as leakage (section 7.6) from pipes which have fractured or internal pump faults, from a blocked inlet to a pump (or air entering) or from a change in fluid condition, e.g. viscosity. It may rise, for instance, because of blockage or jamming of a relief valve. It is a valuable, yet simple, parameter for monitoring fluid power and process systems. Note, however, that the change in pulsations due to partial deterioration of a piston, vane or gear pump must be determined by dynamic monitoring such as described in section 7.2, and care must be taken to ensure that, in this case, the sensor has a wide frequency band. It is possible to have a virtually flat response (i.e. consistent sensitivity) over a very wide frequency range, e.g. 90 kHz for miniature devices.

Table 7.21 Pressure measuring methods

Method	Range (examples)	Accuracy (examples)	Comments
Bourdon dial	From 15 mbar to 8000 bar	From ±5% FS to ±1%	100 bar over-pressure possible Test gauges to a higher accuracy, e.g. to better than ±0.1%
Bellows (or diaphragm)	To 350 bar or higher	To ±0.2% FS hystereris and repeatability	Diaphragm movement may be detected by a capacitance change Mainly process systems Long-term stability
Foil gauge sensor	0 to 500 bar	To ±0.25% FS linearity and hysteresis	High stability Low output
Manometer	Low pressures only	±0.5% to ±0.2% of reading	Glass may be a hazard
Strain–force detector – bonded strain gauge	0 to 700 bar	To ±0.1% FS linearity, hysteresis and repeatability	Gradually being superseded by the types shown below Not suitable above 120°C
Strain–force detector – piezo-electric sensor	0 to 1700 bar	To ±0.05% FS linearity and hysteresis	High stability High accuracy Resists shock Requires charge amplifier
Strain–force detector – piezo-resistive sensor	0 to 1700 bar	To ±0.05% FS linearity and hysteresis	Detects both alternating and steady pressures High output

Table 7.21 shows a range of pressure sensor technology. Note that pressure may either be 'gauge' or 'absolute'; the 'absolute' also includes the atmospheric pressure. In other words, all pressure sensors detect the

difference in pressure between two places, as shown by the three examples below:

1. differential (PD), pressures taken from two places in the system, e.g. either side of a filter;
2. gauge (PG), pressure is taken from the system on one side and to the atmosphere on the other;
3. absolute (PA), pressure is taken from the system on one side and connected to a vacuum pump on the other side (or is sealed with a vacuum).

Accuracy is usually quoted in percentage full scale (FS); on rare occasions a three-figure accuracy is given such as ±3–2–3% meaning that the accuracy of the lowest 25% of the scale is ±3%, over the middle half is ±2% and over the top 25% of the scale it is back to ±3%. Accuracy really only applies to the final display, but for the gauge or sensor alone it is usual to refer to 'non-linearity and hysteresis' and 'repeatability'. The important point is that an adjustment can be made to the absolute value displayed (at a setting point); thereafter the displayed readings must be consistent over the operating range.

The strain–force detectors basically use a diaphragm on which a four-active-element Wheatstone bridge is attached. Pressure applied to the

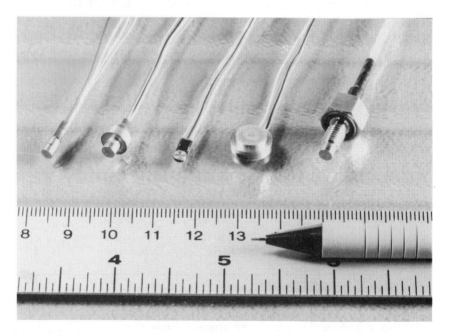

Figure 7.54 Miniature pressure sensors (Entran Ltd).

diaphragm causes it to deflect and hence to produce a signal (compensated for temperature). However, there are significant differences between the different designs. For instance the diaphragm could be made of silicon, stainless steel etc. The bridge can be wire wound, semiconductor, thin film laser-trimmed resistors, piezo-electric or piezo-resistive. The way it is bonded varies from various adhesives to being diffused integrally into the silicon chip. Chemical vapour deposition (CVD) is used to produce high volume low cost sensors with high reliability and good accuracy.

Two, three or four wire connections are made. Two-wire transmitters are preferred to three- or four-wire when the transducer is required to be a long distance from the display or control monitor. Although the two wires provide the transducer power, the pressure value is indicated by modulation of the current (e.g. 4–20 mA).

An example of miniature pressure sensors able to operate up to 350 bar is shown in Fig. 7.54.

(b) Flow

Flow changes of the fluids occur for very similar reasons to that of pressure changes (section 7.9.4(a)). However, although not all flow monitors can be fitted with ease, there are many occasions where such a monitor is preferred, e.g. where pressure compensation is automatically applied in the system and pressure sensing would probably be meaningless.

Manufacturers vary considerably in the accuracy of products. Accuracy and repeatability are not always quoted in the same way – watch out for whether the accuracy is over the whole range (i.e. of reading), or is just stated for the maximum flow (FS). In some cases accuracy may be quoted in a dual manner, e.g. ±2.0% reading +0.5% FS. There is considerable variation in all these devices and effects such as flow pulses could cause a problem. Fluid viscosity and pressure must be taken into consideration with some techniques. All the devices, except the non-intrusive, will cause a disturbance to the flow with the resulting pressure drop. Note that the ranges and accuracies in Table 7.22 are examples.

It will be apparent that the choice of a flow monitor is not easy, and the table only mentions the more common – it has been suggested that there are actually over 200 different types available. (Recognized meter types are reviewed in BS 7405:1991, section 3.3.) Probably some will be discarded because of size or difficulty of fitting or adverse pressure drop in the system. There then remains the debate of cost versus accuracy, or even the performance. Life may enter the discussion where the advantage of no moving parts may be essential. Mass, rather than volume

Table 7.22 Flow measuring methods (liquids)

Method	Range (examples only)	Accuracy (%) At reading	Accuracy (%) At full scale	Comments
Coriolis mass (vibrating tube) (Fig. 7.55)	100:1 1 to 80000 kg h^{-1}	0.15 to 0.5		Virtually unaffected by other physical parameters No R$_e$ restrictions Good for process fluids and food industry Disadvantage: air may cause problem
Differential pressure – orifice plate	3:1		1 or 2	Good for low viscosity fluids Variable advantage: can be improved with quality differential pressure transmitters, e.g. up to 10:1 range
Differential pressure – averaging pitot (Fig. 7.56)	5:1		1 to 2	Simple insertion fitting Low cost High pressures and temperatures Variable advantage: only measures flow at pitot hole; an array is fitted to provide an average flow profile
Electro-magnetic (Fig 7.56)	100:1 or more True flow volume From 0.1 m s^{-1} to 10 m s^{-1} full scale values	0.5+, i.e. ±0.001 m s^{-1}	0.05, i.e. both percentages apply	Ultralow to ultrahigh flows Variable advantage: conducting fluid only
Gear (PD)	100:1 1 mL min^{-1} to 1000 L min^{-1} and above	0.15		Good for viscous liquids
Helical screw (PD)	100:1 0.5 to 2000 L min^{-1}	0.15		Excellent for viscous liquids Up to 250 mm diameter pipe
Nutating disc (PD)	10 to 300 L min^{-1}	2		Low cost Good for water batching
Oval gear wheel (PD)	30 ml min^{-1} to 16000 L min^{-1}	0.15		Excellent meter for chemicals and batching
Paddle wheel	10:1 Velocity from 0.1 to 15 m s^{-1}		2	Low cost insertion type Tolerant of particles Variable advantage: measures flow velocity in region of wheel

Table 7.22 contd

Method	Range (examples only)	Accuracy (%) At reading	At full scale	Comments
Pelton wheel (Hall effect)	10:1 2mL min^{-1} to 20000L min^{-1}	0.3		Ideal for ultralow flows Variable advantage: non-linear
Piston (PD)	1000:1 1.0mL min^{-1} upwards	0.2		Good for viscous fluids Very accurate at low flow for petrol
Rotating vane (PD)	Up to 75mm diameter pipe	0.15		Very high accuracy Petrol pumps and aircraft
Turbine (IR, magnet or Hall effect detection) (Fig 7.56)	6mm to 600mm diameter	0.5		Low viscosities preferable Fiscal meter Disadvantage: damaged by particles >100μm, and needs a strainer
Ultrasonic – clamp on (Doppler)	Velocity from 0.02m s^{-1} upwards		2	Easy to clamp onto pipe Disadvantage: fluid needs bubbles or particles >30μm
Ultrasonic – clamp on (transit time) (Fig. 7.57)	Almost zero to ±15m s^{-1}	2		Independent of conductivity Can be retrofitted Variable advantage: best with flows above 0.1 m s^{-1}; temperature and pressure may affect the result
Ultrasonic – dedicated	See above	1		See above
Variable area (rotameter)	10:1 0.2mL min^{-1} to 750L min^{-1} To 75mm bore		1–6	Low cost for small bore pipes (up to 75mm)
Vortex shedding	10:1 5L min^{-1} upwards 25mm to 0.8m full bore, but can be inserted	1		Good repeatability High pressures No moving parts Variable advantages: pressure drop may be low; best with relatively clean low viscosity fluids' R_e >5000

PD, positive displacement.

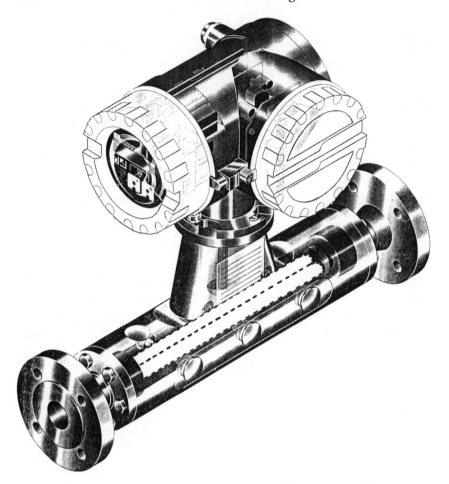

Figure 7.55 Promass two-vibrating-tube Coriolis flowmeter (tubes shown by broken lines) (Endress & Hauser).

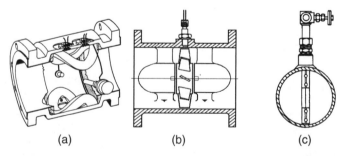

Figure 7.56 Some flowmeter designs: (a) electromagnetic; (b) turbine; (c) Pitot.

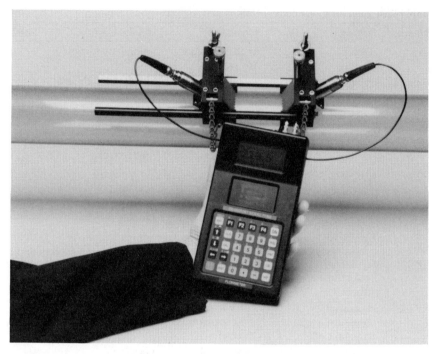

Figure 7.57 Transport PT868 portable hand-held transit time flowmeter (Panametrics).

Table 7.23 Questions and answers for the correct selection for a flowmeter

• What is the line size?	Above 450 mm consider insertion meters.
• What temperatures and pressures are present?	Extremes limit the choice.
• Is the head loss important?	Check the differential pressure.
• What range of flow has to be covered?	Above 6:1 may rule out dp meters.
• What accuracy is needed to detect a change due to a machine fault?	See text below.
• What electrical output is best?	Depending on the data acquisition operated.
• How much money is available for the application?	Costs vary considerably.

may be a requirement. David Gerrard (Flowline Manufacturing) has laid out a series of questions and answers which aid the correct selection for a flowmeter; these are shown in Table 7.23, slightly modified with the machine condition monitoring in mind.

For the machine condition monitoring application, it is essential to be able to measure the flow with sufficient accuracy (or, at least, repeatability

Table 7.24 Level measuring methods

Number	Technique	Type[a]	Advantages	Disadvantages	Cost[b]
1	Acoustic cavity	C	Small intrusion Any attitude	Still relatively new	?
2	Bubbler pressure	C	Simple Churn immunity	Needs good air seal Limited attitude	c–h
3	Capacitance	CS	Simple Coating immunity	Fluid sensitive Special cabling	d–h
4	Conductance	S	Simple Small probe	Conducting fluid only Froth effects	c–e
5	Displacement	CS	Low cost Quite simple	Vertical only Sticking possible	a–i
6	Float	CS	Low cost Simple	Moving parts Limited range	a–f
7	γ radiation	CS	Non-contacting High penetration	Expensive Large	g–i
8	Hydrostatic pressure – direct	CS	Slurry resistant Quite simple	Sensitivity problem Temperature limitation	c–g
9	Hydrostatic pressure – transmitted	C	Slurry resistant Many types available	Relatively expensive Sensitivity problem	e–h
10	Indicator tube	C	Clear visual Simple	Local May be large	f–h
11	Load cell	C	True bulk External	Fitting difficulty Vulnerable	?
12	Neutron scatter	S	Non-contacting 100 mm+ walls	Wall-hydrogen effects Expensive	i
13	Optical refraction	S	Very simple Small	Coating effect Temperature limitations	b–h
14	Radar	C	Narrow aperture Good accuracy	Sophisticated Expensive	i
15	Resistivity	S	Water–steam–air High pressure	Possible debris effects Expensive	h
16	RF admittance	S	Coating immunity No maintenance	Some interaction Setting-up difficulties	ef
17	Rotary paddle	S	Liquids and solids Totally enclosed	Moving parts Large	de
18	Sight glass	C	Low cost Very reliable	Local Fragile	b–e
19	Thermistor	S	Small and simple Multi-level possible	Debris effects High probe temperature	cd
20	Time domain reflectometry	C	Small intrusion Flexible sensor	Still relatively new	?

Table 7.24 contd

Number	Technique	Type[a]	Advantages	Disadvantages	Cost[b]
21	Ultrasonic – wet	CS	Small (or long) Coating resistant	Aeration effects Multiple echoes	d–f
22	Ultrasonic – dry	CS	Non-contacting Compact	Wall thickness Temperature sensitive	f–i
23	Vibrating tube or tuning fork	CS	Coasting immunity Temperature range	Tube affected by vibration and aeration	d–g

[a]C, continuous; S, switched.
[b]Cost categories in 1995; a, £1–5; b, £5–20; c, £20–50; d, £50–100; e, £100–200; f, £200–500; g, £500–1,000; h, £1,000–2,500; i, over £2,500.

from day to day) so that the slight changes due to a fault developing will be detected at an early stage. The difference in flow for the expected fault should be calculated and then compared with the type of accuracy possible from the different monitors as given in Table 7.22.

Air flow is discussed in section 6.1.

(c) Level

The reservoir level of oil or process fluid not only gives an overall indication of supply and demand but also conveys whether there are faults within the system. In other words, not all the fluid being 'used' is providing the required function, but is being lost by external leakage. Note that this monitoring does not cover 'internal' leakage; that must be assessed from a knowledge of the efficiency of the system.

Level detectors are either continuous (C) or one level only (i.e. acting as a switch (S)). Table 7.24 briefly lists 23 different techniques for measuring level, and Fig. 7.58 gives a simple illustration of each of them fitted to a reservoir. At 1995, the type of cost associated with the different techniques is approximately described as summarized in Table 7.24.

(d) Speed (rotational)

Speed can be related to the linear motion of some component or the rotational speed of a spindle or axle. Linear motion is detectable by such as an LVDT (linear variable differential transformer). This section, however, only discusses the rotational application. Quite frequently a vibration will occur in the rotary sense; this is discussed in section 7.2.

Rotational speed is of major importance, and any drop or rise beyond the acceptable limits will indicate either a serious fault in the machine or that a serious problem will shortly arise.

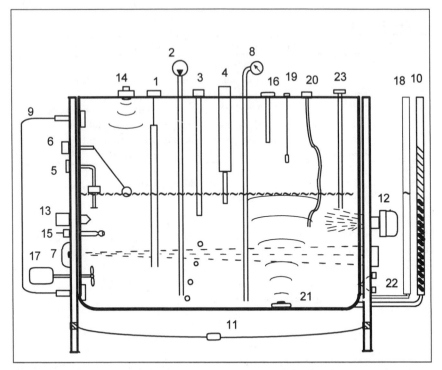

Figure 7.58 Examples of the fitting to a reservoir of the level sensors (1–23) in Table 7.24

Rotational speed may either be sensed purely as a number of rotations (per minute, say) or as a constant rate of rotation. The latter may vary during a single rotation, and for this an 'encoder' is used as an axially attached unit with a highly accurate series of line graduations on glass as the measuring standard; an LED–photodiode is used to sense these lines.

Methods for measuring speed are listed in Table 7.25.

(e) Temperature

To some extent temperature has already been described in section 7:6. However in that section the important aspect was the temperature distribution or pattern. This section now deals with individual spot measurement of temperature – mainly by contact but also by radiation. (An example of radiation pyrometry monitoring is in the examination of turbine blades while the turbine is operating, which can convey something of the efficiency of the total engine.)

Table 7.25 Speed measuring methods

Type	Range	Comments
Pick-up (magnetic etc. (Fig. 7.59)	0.001 rev min^{-1} to 300 000 rev min^{-1}	Simple metal projection attached to or part of rotating shaft and sensed by a magnetic pick-up or proximity sensor (usually once per revolution)
Stroboscope	Relatively low speeds	Strobe adjusted to just give a stationary image of the shaft, i.e. once per revolution
Hand-held – contact	3 rev min^{-1} to 100 000 rev min^{-1}	Tachometer pressed on to shaft axis; not suitable where the axis is inaccessible
Hand-held – remote	3 rev min^{-1} to 500 000 rev min^{-1}	Optical beam (or IR) projected up to 1 m away from the shaft either on to the circumference or axis end
Encoder – optical	Up to 5000 lines per revolution to 200 kHz (i.e. 2400 rev min^{-1})	This is more appropriate for sensing the position of rotation rather than the revolutions per minute, i.e. the uniform rotation
Electric motor magnetic disc	Up to 25 kHz	A small box houses a complex circuit which senses the pulses from the magnets for comparison with preset maximum and minimum settings
Electromagnetic torque	From 15 rev min^{-1} to 1000 rev min^{-1}	Used to activate switches when the speed reaches a certain amount (torque); they work in either direction of rotation

Temperature is normally quoted in 'degrees Celsius' (quite frequently spoken of as 'degrees centigrade'). Absolute temperature is the temperature above the lowest temperature theoretically possible, i.e. $-273.15°C$. The kelvin scale uses this minimum temperature as zero, with the same scale interval; thus

$$0°C \approx 273K, \ 100°C \approx 373K.$$

Temperature can be measured in numerous ways although the thermocouple and resistance thermometer are the most common in machinery. Table 7.26 shows the range with an idea of the temperature scales possible. (Others not shown are the gas pressure type, electrical noise, acoustic transmission and total radiation, which either are rarely used or are extremely complex and expensive.)

Figure 7.59 A speed sensor in position (Ringspann).

The choice of temperature sensor is influenced by its requirement. Initially in setting up a development system it will be important to know the absolute temperature with good accuracy, particularly if standards are to be defined. However, thereafter for condition monitoring in real applications it will be the trend that is required, i.e. a good repeatability. So while resistance thermometers may be preferred for the development, the cheaper thermocouple may be used for the real process. A range of temperature sensors is shown in Fig. 7.60.

Table 7.26 Temperature measuring methods

Method	Range	Comments	
		Positive	*Negative*
Mercury in glass	–36°C to +500°C	Good visual indication High accuracy	Fragile; not suited for remote monitoring
Mercury in steel	–40°C to +650°C	Suitable in hazardous areas	Distance between sensor and actuator <2m
Thermocouple	–250°C to +2000°C and even higher in some cases	Widest temperature range; simple application; robust Low cost and small size	Needs a temperature reference; not easily suited to high accuracy except for <100°C difference between cold and hot junctions Special extension cables needed for long runs Susceptible to shock
Resistance – wire wound platinum	–230°C to +850°C	High accuracy is possible; simple installation; copper cable only needed	Needs energizing source Careful handling needed; larger than thermo-couples Slower response than thermocouples and thermistor
Resistance – platinum film on ceramic substrate	Up to +500°C	Cheaper than wire wound; good thermal contact with flat surfaces Vibration resistant	Upper temperature limitations Needs protection from environment
Thermistor	–40°C to +300°C	High accuracy over small range, owing to high output; changes as small as 0.0005°C can be detected with accuracy of ±0.2°C Shock proof Small size	Limited range
Infra-red (radiation pyrometry	–55°C to 1350°C although band may only be 400°C	Remote (up to 40m)	Surface emissivity needs to be known; narrow receptance angle required
Bimetal	–70°C to +540°C	Low cost and robust Safe in hazardous areas	General purpose only Limited accuracy and slow response Best for liquids

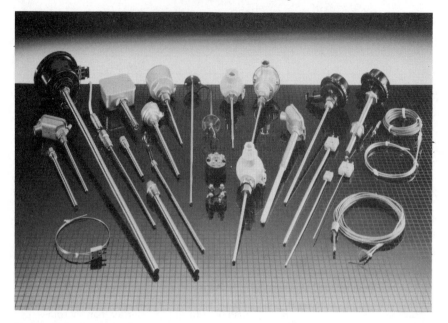

Figure 7.60 A range of temperature sensors (Labfacility).

Speed of response may be important where there are fluctuating conditions. The very small temperature changes which may need to be known are best undertaken by thermistor.

7.9.5 How to start

Having decided what simple single parameter conveys a condition of the machine, examine the lists given above. If it is not one of the five examples given, then approach several suppliers and ask for their comments along similar lines to the examples here. Get a feel of what is available and try two or three *in situ* on the machine.

It must be remembered that the environment and adjacent machinery can affect gauges, e.g. by vibration or temperature.

The output needs then to be logged at regular repeatable occasions – Chapter 9 – and a feel achieved of the acceptable levels. Going outside this 'acceptability' will signal a problem arising.

CHAPTER 8

Output monitoring

To monitor the output of a machine is to look at what it is producing. It may be the required product, it may be the by-product or it may, however, be signals of distress. In addition, the total ability of the machine to operate successfully is highlighted by how effectively it can produce – its performance. Some of these features will be examined in this chapter, and various means of their monitoring will be discussed. Individual monitors cover a wide range of ideas but only those appropriate to the process in hand will need to be considered before choosing one; however, to give an idea of what is possible, some of the more specialist monitors are described in this chapter.

8.1 PERFORMANCE

8.1.1 Introduction

The measurement of the performance of a machine is perhaps the most obvious of condition monitoring methods. After all, the machine is meant to produce a satisfactory product or service, in a satisfactory time, and if that satisfaction reduces then it is likely that a fault of some sort has developed. Performance is the overall way a machine is operating; it looks at all the facets of the operation and assesses whether the machine is at its optimum condition.

The general concept of performance is more than just output satisfaction, in that it covers the behaviour of the machine in all its features. For instance, performance could include the following:

- speed (or response) of operation – is it becoming slower (or faster)?
- ease of manipulating controls – is it becoming stiffer (or looser)?

Performance monitoring thus covers both how well the machine is working and how well it produces an output. In effect, performance monitoring is efficiency monitoring. (Other features also come into the efficiency equation, e.g. noise and temperature, because should they change it may be because some proportion of the power is being used in their generation.)

In order to be able to determine the efficiency we need to know both the input energy absorbed and the output work done, or, for a constant input, to know the output.

8.1.2 Brief description

What you get for what you put in may well reveal what condition the machine is in. In addition the effluent is an indicator.

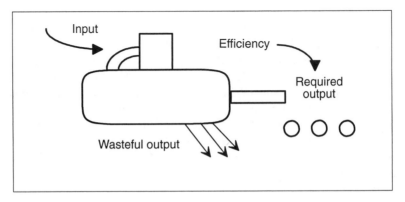

Figure 8.1 Output from a machine.

(a) Efficiency

No machine is 100% efficient. There is always some loss or decay. The second law of thermodynamics reminds us that since the original creation of the world available energy is decreasing unless there is an additional input from an outside source. The loss of useful energy in a machine operation will be due primarily to friction and temperature effects, and if these increase, owing to a malfunction growing, then the efficiency will decrease:

$$\text{Efficiency} = \frac{\text{useful work done by the machine}}{\text{total energy used in driving the machine}}.$$

(b) Output

Changes (or increases) in the power input to achieve a constant output have already been discussed in section 6.2. That may well be the way the machinery is designed and hence the better way of monitoring performance. However, the majority of machines have a constant input and the deterioration is shown in the changing (or falling) output.

8.1.3 Practical applications

How efficiency is determined will vary with each machine. The input and the output need to be 'measured' at a specific time after start-up and during a specific mode of operation which repeats each day, i.e. the efficiency must be measured and compared for the same conditions. Take the examples given for the machines mentioned in Table 8.1.

Table 8.1 Efficiency measurement from input–output

Type of machine	Input measured (Chapter 6)	Output measured
Machine operation	Electrical power consumption	Physical work done measured by a load cell and displacement
Gear box	Torque and speed applied to input shaft	Torque and speed at the output (section 7.9)
Hydraulic system	Electrical power consumption in a set time	Work done in the same specific time
Hydraulic pump	Electrical power consumption	Pressure and flow produced

Figure 8.2 shows the change in efficiency of a gear pump as it wears. Its efficiency varies with the pressure being applied, so the graph represents efficiencies over a range of pressures. In this case there is the added indication of temperature, as discussed in section 8.1.4.

8.1.4 Detailed discussion of machine output measurements

One method used for determining the efficiency in Fig. 8.2 was that of thermodynamic efficiency. Thermodynamic efficiency is now a recognized technique for the measurement of hydraulic pumps, particularly of water pumps but also of other types. In simple terms, it can be considered as the change in temperature of the liquid as it passes through the pump – due to power losses within the pump. So if the pump begins to wear and to become less efficient, the temperature rise will increase. The temperature will also rise as a result of the change in pressure, so the

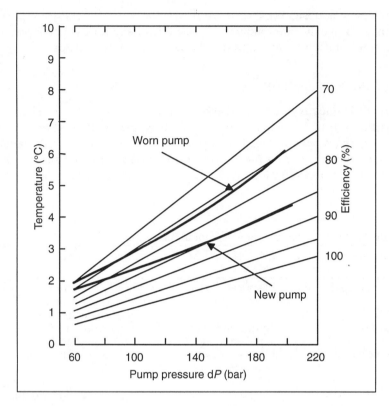

Figure 8.2 Efficiency change due to wear in a gear pump (Witt *et al.*, 1977).

detection device needs to see the additional change. Figure 8.3 shows an example of a monitor used for water pump monitoring. (Section 13.10 should also be referred to.)

The monitoring of electrical current being produced by an electricity generator is a good example of output monitoring on its own. The generator is being driven by a constant speed turbine, e.g. water, steam, gas, diesel etc., and so the output change is related solely to the generator (unless, of course, there is a major failure of the constant speed turbine, which is monitored separately).

8.1.5 How to start

It is first necessary to decide how the machine or system can be assessed as regards its performance. This may be quite simple if it is already being undertaken. However, there are many systems which have not had the luxury of being so assessed; these need considering from scratch.

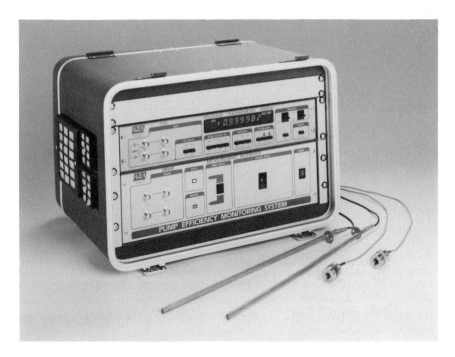

Figure 8.3 Thermodynamic efficiency water pump monitor (ASL).

- What is the power source?
- How can it be assessed accurately?
- In what way can the output be assessed with precision?

8.2 PRODUCT VARIATION

8.2.1 Introduction

Product variation is indicated by the usual inspection programme. For instance, if the product has to be of a certain size, then the size is periodically measured (sampled) and if it goes out of the acceptable range (above or below) then the machine needs a service of some sort. It may be necessary to tighten a certain part or adjust the setting; on the other hand, it may be wear in a machine operation part which necessitates a replacement.

8.2.2 Brief description

The monitoring requires the output of the machine, in the sense of the product produced, to be examined for unusual flaws and significant deviations.

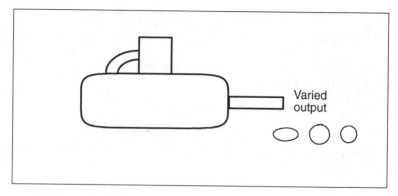

Figure 8.4 Output variations.

8.2.3 Practical applications

Any production machine or system can be monitored in this way, although the detail which can be found will vary considerably depending on the product. Table 8.2 gives some ideas of the type of successful monitoring which could be undertaken.

Table 8.2 Flaws in finished products

Product	Machine fault	Indication
Any product requiring a clearly defined fit	Loose or worn bearings causing poor alignment	Size of one or more pieces is not matched to its connector
Any sheet metal object with a smooth surface finish	Fatigue of cutting edges causing pieces of debris to get into the coolant of the punch or stamping machine	Indentations on the surface, or streaks if using a sweep motion Gloss monitor may be used
Any product which must appear desirable to a customer	Wrapping or printing machines, or those involved in producing clearly defined shapes, outputting an unacceptable result	Dimensional differences or change in outline Colour sensing possible

8.2.4 Detailed discussion

It is important that product variation is not just seen as a production feature; it needs to be included in the overall monitoring programme. A product deficiency may be the first signs of a serious mechanical fault. The fault may be striations or the occasional inclusion or indentation on a surface, as well as the more obvious size and shape deviations.

A periodic fault shown up by every fifth component, say, may be a fault developing on one gear in a gear train, or it may go back even further to a fault on the gear hobbing machine which influences the characteristics of the gear wheels it produces. (However, in the machine case this can be traced and put right after the monitoring has shown it up as an important feature.)

8.2.5 How to start

This is a fairly remote type of monitoring. However, it should not be totally disregarded and machine operators, for one, should be aware of its importance in the visual sense, if nothing else. If surface finish is important, then the use of a 'gloss meter' may be an ideal and simple first type of monitor. Infra-red or image analysis dimensional checking may be undertaken to detect changes in dimension and shape.

8.3 SIGNALS DOWNSTREAM

8.3.1 Introduction

In some ways, the signals downstream are the same as the product variation – they are what is issuing from the machine. However, while the product, which is the purpose of the machinery, has to be often checked for correctness by production inspection, other signals, which are probably the 'effluent' as it were, could be the exclusive right of the monitoring engineer.

8.3.2 Brief description

Where there is a flow of something to waste from a system, or where there is a flow output from a component, the system or component can be monitored by looking at the flow as it is emitted. This may involve a detailed examination of the liquid or gas or solids involved, or it could be a total look at the fluid such as its flow rate.

8.3.3 Practical applications

Any system or component which involves a process or effluent flow output. Table 8.3 gives a couple of ideas.

Table 8.3 Flaws showing in fluid output

System or component	Machine fault	Indication
Pump	Leakage in one cylinder	Irregular pressure fluctuations measured by pressure transducer in the process liquid
Automobile	Cylinder wear Blockage of air filter Deterioration of carburation setting	Excessive hydrocarbons in the exhaust Excessive oil fumes

8.3.4 Detailed discussion

Wear debris analysis, described in detail in section 7.4, can also be applied to the output of a machine. In other words, the process fluid from the machine may contain debris which originated from a wearing component. Although the debris analysis is best conducted on the liquid withdrawn from within a machine, uncontaminated from any external fittings, it may not always be possible to take a sample within; an external flow is then the only option available.

In addition to the debris, other valuable signals can be seen in the downstream fluid; these are described in section 7.5 (Table 7.11). There is also the gas, where this is part of the process, e.g. the flue gas. Both particulates and gases can be monitored in the emission stack or exhaust system. An example is that of the Erwin Sick flue gas monitor, the basic idea of the working of which is shown in Fig. 8.5. The optics is very complex but it enables SO_2, NH_3 and NO to be continuously monitored and a measure of the particulate above $5\,mg\,m^{-3}$ to be detected also.

Smaller concentrations of particulate gases can also be detected using techniques very similar to the optical devices for looking at individual particles in liquids (section 7.4). A completely different technique, detecting a very wide range from something like $0.1\,mg\,m^{-3}$ to $1.2\,kg\,m^{-3}$, is covered by an electrostatic technique developed by the Wolfson Centre for Bulk Solids Handling Technology in Woolwich.

The range of possibilities for the monitoring of gas (and smells) which are emitted by a machine is given in tabular form in Table 7.3. In particular the flue gas is likely to be monitored by solid electrolyte, such as the on-line zirconium oxide oxygen sensor first introduced in 1971 by Rosemount Limited: other types include chemiluminescence, FTIR, NDIR, UV, paramagnetic and crystal based etalon.

Because of the large market in automobiles, vehicle exhaust monitoring probably accounts for around 50% of the pollution monitoring of the atmosphere, and this can be associated with both the quantity of vehicles operating and their condition.

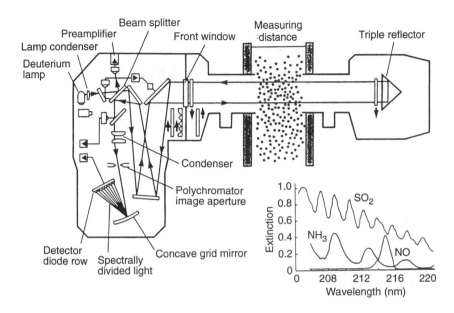

Figure 8.5 The detection and analysis of flue gas (*Opto & Laser Europe*, May 1995 p. 35)

8.3.5 How to start

Tables 8.3 and 7.11 will give some ideas of possible monitoring, but there is a need to consider thoroughly what is issuing from the machine – both intended product and expected effluent.

CHAPTER 9

Data acquisition

We have discussed numerous means of sensing in the previous chapters. The right and apt sensor is critical for reliable decision making. However, just as important is the means by which that sensor information is acquired and passed to the 'decision maker'. If there are hidden bugs which totally prevent the transfer of data, so be it, but if they marginally influence helpful data on the way to the analyser, we will have misleading results, and we may well be in for unnecessary trouble.

The sampling of the sensor output may involve numerical data logging, or it may require suitable hardware to transfer oil or high frequency spectra. Each of these will be discussed here, as well as the advanced means of remote communications. Data acquisition is a field which has made rapid progress in just a few years owing to the enormous expansion in computing capabilities; this applies not only to the quantity of data which can be stored in a small computer 'chip' but also the rate at which the data can be acquired and transmitted.

9.1 WRITTEN OR PRINTED REPORTS

The logging of data may be undertaken by hand, literally. Hand-written detail on carefully designed forms is no bad thing in certain circumstances, but we must remember that the data on the form still need to be taken off in some way in order for the analysis to be undertaken. However, the accuracy of the data, these original data, is probably quite high, and certainly higher than data punched in on a terminal or keyboard. There is another advantage in that the person who obtained the data is usually easily identifiable.

Where the person involved is reasonably intelligent, or at least reasonably experienced, they may well be able to notice a faulty reading instantly. They may even be able to detect a fault developing in the system; although that is a bonus it is the concept of accuracy of data

HYDRAULIC SYSTEM INSPECTION	Ref.		Week No	
Location				

PRESSURE

		bar	Alarm ✦
	TP1		
	TP2		
	TP3		
	TP4		

RESERVOIR

Level (mark with an →)		Normal / Low	Temp. at Time	°C am/pm	Alarm ✦

✦ Tick if Alarm is indicated

FILTRATION

Tick in one column only		GREEN OK		AMBER change		RED stop	
	F 1						
	F 2						
	F 3						
	F 4						
	F 5						

OBSERVATION

Leakage	
General Appear-ance	
Noise	
Smell	
Perform-ance	

Readings taken on Date		Name		Checked	

ACTION NECESSARY		To:
	Completed Date	Name

Figure 9.1 Example of a hydraulic system data sheet.

which needs to be addressed here. A reaction such as 'That's impossible!' or 'I need to check that again' will save much time later when eventually the data are officially analysed.

In order to retain this accuracy and to prevent data being lost (as can happen with loose-leaf filing) it is preferable to use pre-printed hard-back books with already numbered pages. If a mistake is made cross through the page, but do not remove it. Each page (or record) should contain as a minimum the following background information:

- date;
- time;
- place and machine;
- position of connection of monitor or sampler;
- person logging the data.

An example is shown in Fig. 9.1.

Written reports also need to have a further block included indicating the action taken, i.e.

- how and when and by whom the data were transposed for analysis.

If this is not completed, then a later examination of the paperwork will indicate that nothing has happened.

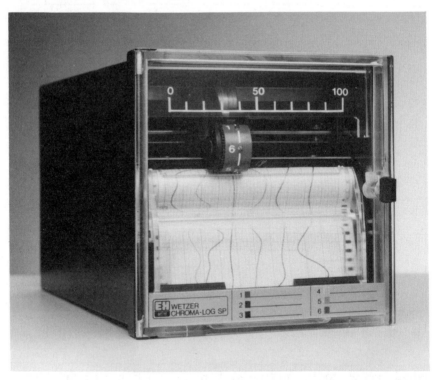

Figure 9.2 A six-channel colour chart recorder (Chroma-log – Endress + Hauser).

It goes without saying that the use of such written reports implies that there is a process whereby they are used. Thus a maintenance programme must include a daily, or weekly, routine examination and logging of the situation, followed by a specific time for the data to be processed. It must be decided whether the 'book' itself can be removed from its usual place or the important 'information' only is taken (by photocopying or transposing to a separate sheet).

Written reports only refer to steady state conditions, periodically examined. Where there is a possibility of a continuous gradual change occurring, or where a dynamic signal needs recording, then a chart recorder may be attached to the system. Figure 9.2 shows an example of a four-colour chart recorder. The natural progression from the hand-written data sheet is the computerized logger with a keyboard and, possibly, an interface to the sensor.

9.2 Portable loggers

The multipurpose portable logger normally appears to be little more than a portable keyboard allowing the maintenance personnel to log the sensors at regular intervals directly into a computer form, rather than a

Figure 9.3 Example of a portable logger (Mon-a-log – Cambridge Monitoring Systems).

hand-written form (Fig. 9.3). It can store many more report forms than a normal book of similar size could contain. As mentioned above, the data may not be quite as consistent as the hand-written form owing to typing mistakes, so care must be taken to back check every so often to see whether the information is correct and is correctly stored.

A more advanced portable logger will overcome the problems associated with personnel wrongly punching the data in; it does this by the use of an interface directly between the sensor and the hand-held logger store. In this case the interface may be within the logger itself or it may be a separate unit which is inserted in the line, or it may be part of the sensor hardware. The portable instrumentation tape recorder could be mentioned here where very high precision can be achieved from 0 Hz to 20 kHz for say 4 channels, with a lowering of the range depending on the number of channels (to, say, 2.5 kHz at 64 channels).

These loggers are often pre-programmed to obtain the necessary data and, for the punched-in data, will provide prompts to the user. In some instances a 'default' value may be suggested in order to cover all the data requirements in as short a time as possible. This is dangerous. It is too easy for the personnel to blindly go through, pressing the 'yes' button, when the true data may in fact be available. In no way may critical data be allowed to be given a 'default' value, or if it is this should be made clear in the final result.

The frequency of logging may not be important where a manual input is operated. However, to be able to plug in the unit and to leave it to sample data automatically every so often can have considerable advantage – the unexpected will not be missed. At least one unit on the market can be pre-programmed to sample at different rates over a period as long as 6 months.

The output from a portable logger may be

- on a miniature screen,
- on a paper printer (roll) or
- downloaded to a desk-top computer.

In the case of the screen display and the paper roll output, the room available is very limited, so care must be taken to ensure that only what is essential is output. The built-in software should allow much more data to be scrolled across (or up and down) the miniature VDU screen but this is not always convenient. The full data are usually still left in the logger store for examination or downloading at a later occasion. In addition it may be possible to have a

- basic alarm level set

such that either a light comes on or a buzzer sounds if the alarm level is exceeded.

9.3 ON LINE TO COMPUTERS

This is where considerable improvement has been made in the last few years. There is no need for down-loading from a portable; the data are automatically transmitted to the PC from the sensors (via a suitable interface). The restriction in the past has been the maximum permissible rate of transfer, but with PCI (Peripheral Component Interconnect) the PC is no longer the restriction and can cope with around 1 MHz data for display. The interface card standard, covered by the PCMCIA, is little more than the size of a credit card but is able to provide the transfer necessary.

The objective here is to be able to use the computer not only to monitor the critical functions of a mechanical system (e.g. output), but also to monitor the condition of the machinery through appropriate sensors (as dealt with in Chapters 6–8). The idea of remote control rooms is nothing new, but it is the specific remote monitoring of the 'condition' function which is the new feature.

This is where the more sophisticated term 'data acquisition' is commonly used. The interface may be a separate unit or a plug-in 'board' with the electronics and memory and software all included in

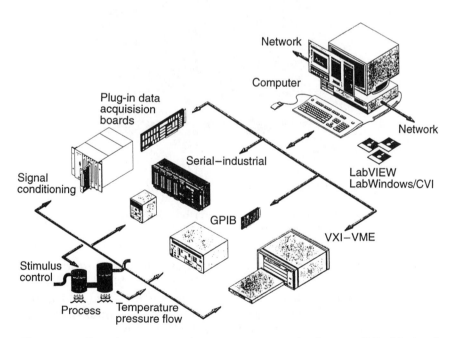

Figure 9.4 Interface routes from process monitoring to PC (National Instruments).

one piece of computer 'hardware'. The price of these boards can vary considerably depending on what functions they include, from hundreds to thousands of pounds sterling. Figure 9.4 shows a range of input interfaces supplied by National Instruments, and the necessary software for the PC; in this case the PC becomes a series of 'virtual instruments', such as 'virtual scope', 'virtual spectrum analyser', etc., as well as effectively displaying the sensor values. This has many advantages such as low cost, flexibility, speed, multi-input etc.

Figure 9.5 A typical plug-in data acquisition board (plugs into PC bus slot) (Strawberry Tree – Adept Scientific plc).

Figure 9.5 illustrates a typical plug-in board which is fitted inside a PC. In some cases a separate, external, interface would be needed to condition the sensor signal. The signal conditioners would take a low level analogue signal from the sensor, amplify it to a maximum just short of the ADC maximum input and digitize it; this process should increase the resolution and decrease the noise. Further aspects of the digitizer are that of isolating the sensor from the PC and of filtering the signal so that only that part needed to be analysed is fed into the PC board; sensor stimulus is supplied and there may be improved linearity. The method of data acquisition adopted, however, will depend on the type of sensor which is in use. For instance, merely logging a slowly changing pressure will require little store, but where a full frequency spectrum is involved (up to say 110 kHz sampling rate) there will be a need for considerable memory (store). The following are some of the key issues to be decided before choosing a data acquisition system:

- how many channels (1, 2, 100, 1000 etc.);
- speed of sampling;
- resolution of input data (8 bit, 12 bit, 16 bit etc);
- resolution of output data (8 bit, 12 bit, 16 bit etc);
- input–output ports (RS232, or multidrop RS485–RS422);
- parallel or serial port connections;
- analogue input levels;
- signal conditioning required.

The more important, less obvious ones are described below (for further details see Keithley MetraByte's *Reference Guide*). The relationship of bits and bytes is

$$1 \text{ byte (B)} = \text{a group of 8 bits (b)}$$

and baud rate is the data transmission rate usually expressed in bits per second.

9.3.1 Analogue–digital sensor signal

A basic sensor is likely to output an analogue signal, i.e. its signal varies gradually rather than in steps (digital); this may be over a range of voltage (e.g. 0–5 V) or current (4–20 mA) or some other detectable parameter (resistance, frequency etc.) More advanced sensors have built-in AD (analogue-to-digital) converters which output a more stable signal unaffected by temperature and length of line.

9.3.2 Input resolution

The input resolution is usually specified in bits – 8, 10, 12, 14, 16 – with 12 the most common at the current time. The conversion from 'bits of resolution' to actual resolution is found from the equation

$$\text{actual resolution} = \text{one part in } 2^n,$$

where n is the number of bits, which is thus 1 part in 4096 for a 12-bit device. In practice one needs to know the voltage resolution; thus for a 5 V FS (full scale) input range the resolution is

$$5\,\text{V}/4096 \approx 0.001\,22\,\text{V}, \text{ or } 1.22\,\text{mV}.$$

This little calculation is important if the optimum device is to be used with the sensors in the system. If a pressure sensor has only an accuracy of ±1% (1 part in 100) then a 16-bit AD board would be over the top as regards expense.

9.3.3 Input accuracy

Three of the more common means of expressing input accuracy are shown in Table 9.1 (for a 12-bit AD converter and a 10 V, full scale input).

Table 9.1 Means of expressing input accuracy

| Specification | Measurement accuracy | |
	Equation	Result
0.024% of reading ± 1 bit	$10\,\text{V} \times [0.024/100 + 1/2^{12})$	4.8 mV
± 2 bits	$2 \times 10\,\text{V}/2^{12}$	4.8 mV
4.8 mV	$10\,\text{V} \times 0.048/100$	4.8 mV

9.3.4 Maximum sampling rate

This must be able to cope with the rate of signal change which is being input. While the maximum sampling rate is often taken as twice the highest frequency component in the input signal, a factor of 3 is really advisable to avoid error. If this is not possible then an anti-aliasing filter is important to prevent unclear (and possibly unwanted) higher frequencies being read as low frequency signals. (Typically, the anti-aliasing filter is set with a cut-off frequency of ⅓ the sampling rate.)

Typically sampling rate is quoted in 'samples per second' (or ksamples per second or Msamples per second), and this is the sampling rate for a single input channel. If more than one channel is being sampled (through a multiplexer, one at a time) then the maximum actual sampling rate is the maximum divided by the number of channels. In practice, the sampling rate may be even less owing to a change in gain associated with each channel.

9.3.5 Serial interface

The three RS serial interface standards are very common. By definition data are transmitted, one bit at a time, over a single communication line to a receiver. It is particularly efficient for sending data at low rates or over low distances.

RS-232 is the oldest and most widely used standard. It now connects two devices, with the transmission line of one device connected to the receiver line of the other. This means that both devices can talk simultaneously (full duplex). There is a maximum recommended spacing of 15 m (50 feet) between connected equipment and a maximum data rate of approximately 2000 characters per second (20 kbaud).

RS-422 is a newer standard and operates over greater distances than RS-232, i.e. to 1220 m (4000 feet) and up to 10 Mbaud (but note that the maximum distance and maximum data rate cannot be applied at the same time). It is also less affected by noisy environments than the RS-232. Although the RS-422 supports full duplex operation, unlike RS-232 it can drive multiple receivers; perhaps as many as 10 devices can communicate. RS-422 hardware is not compatible with RS-232.

RS-485 is the newest standard. It is of similar nature to the RS-422 but it cannot generally be interfaced with it. Because of the better quality, up to 32 devices can share the same connection and the maximum distance and data rate can be achieved at the same time. The different interfaces are compared in Table 9.2.

Table 9.2 Serial interface comparisons

	RS-232C	RS-422A	RS-485
Mode of operation	Single ended	Differential	Differential
Number of drivers and	1 driver	1 driver	32 drivers
receivers allowed on line	1 receiver	10 receivers	32 receivers
Maximum cable length (m (feet))	15.2 (50)	1220 (4000)	1220 (4000)
Maximum data rate (bits s^{-1})	20k	10M	10M

9.3.6 IEEE-488

A further interface of importance is the IEEE-488 interface. This has been a world standard since 1975 and was developed to enable programmable instruments to be attached to computers. The IEEE-488 is also known as GPIB (General Purpose Interface Bus) and HP-IB (Hewlett-Packard designed it in 1965).

9.3.7 Parallel/serial ports

Normally data acquisition is undertaken using serial ports, but terminal block interfaces can be obtained for attaching thermocouples, accelerometers, millivolt input, milliampere input etc. to the parallel port.

9.3.8 Signal conditioning

A separate conditioner can be purchased for each sensor used; however, the more impressive data loggers include an advanced 'universal input' where the logger automatically recognizes what sensor is attached and configures the correct signal conditioning. This will cover such items as accelerometers, resistance thermometers, thermocouples, or any DC or AC voltage inputs.

9.3.9 Analysis

How the data are analysed is dealt with in the next chapter but it is worthwhile anticipating what will be required when a data acquisition system is purchased. Some logger software includes a full graphical package enabling an immediate impression to be gained of what has been (or is) happening in the system being monitored.

9.4 SAMPLING

The sensor may be a highly reliable and accurate device but the value of the analysis depends entirely on whether it is looking at a representative sample from the sensor. Although sampling technique varies from application to application, from type of monitor to type of monitor, there are some basic rules which apply to all. In Chapter 2 we stressed the need for the sampling point to be relevant and convenient. Consider the extension of that in the following.

- The sampling point must be in the correct place, be in the correct direction, allow no interference from other features and permit consistent sampling.
- The sampling time must be when the machine is being operated correctly, be when the machine is thoroughly warmed up and be at the same time after start-up (for comparisons).
- The sampling attachment must be secure.

We have already considered electronic sampling with the various sampling rates in the previous section. Now we review the more practical methods. Because this book will be used as a reference book after a first reading, the following sampling ideas are separated into the relevant types of monitoring described in Chapters 6–8, i.e. those which require 'sampling':

- environment;
- power;

- vibration;
- ultrasonics;
- wear debris analysis;
- oil analysis;
- corrosion;
- product.

In some cases this has already been explained in detail, and hence this will just refer back to the relevant sections; otherwise, the comments are new. Bear in mind, however, the general comments given above.

9.4.1 Environmental sampling (section 6.1)

Most environmental monitoring is directional, i.e. it is influenced by the direction in which the sensor lies. This can be counteracted by having a three-axis monitor (with three sensors) or by a rotating random direction sensor which detects the maximum and minimum (and average).

The logging of such data has, conventionally, been undertaken by sensitive chart plotters, but this is no longer essential. Digital logging, as described above, is likely to be much more accurate, but it may not reveal to the human quite as much as the chart plot. For instance, the trends and spikes on a chart are immediately visible and comprehensible to the human, but are those maxima and minima and progressions going to be analysable from the digital data? Some care will be needed to decide the frequency of digital monitoring so that neither too much data is recorded, nor too little.

9.4.2 Power sampling (section 6.2)

In the case of electrical power, there will be a need to differentiate between surges from normal power supplies and those differences which are exceptional – due to machine faults. In this case it may be possible to isolate, or to reject, 'erroneous' readings at the time of sampling by fitting a mains monitor in addition to the machine power monitor.

Monitoring relating to the use of fuel may be confused if there are deficiencies in the fuel itself. This is unlikely, but it is possible that the presence of water in fuel could be such a problem. A skilled operator should be able to sense such erratic running and to indicate in some way that the machine condition monitoring will be misleading.

9.4.3 Vibration sampling (section 7.2)

There are numerous hints already given in section 7.2, in particular relating to the appropriate equipment and temperatures. However, the major problem is always concerned with what is actually being sampled. In so many cases the sampled signal includes signals from adjacent parts, or even adjacent machinery. There is a need, if possible, to provide damping or isolation of other influences. Another technique is to endeavour to carry out a cross-correlation where, by moving to different positions, the source of the vibration can be positively identified.

9.4.4 Ultrasonic sampling (section 7.3)

Because of the high frequency involved, there are fewer problems with sampling. However, care should be taken to ensure that a representative signal is being achieved by seeing how well the signal is repeatable.

9.4.5 Wear debris analysis sampling (sections 2.2 and 7.4)

Section 2.2 goes into great detail on sampling procedure and techniques. It is highly important that the correct process is followed owing to the ease with which contaminant can enter a sampling pipe or bottle.

9.4.6 Corrosion sampling (section 7.8)

The chief problem with corrosion sampling is the time. Corrosion may be measured in scales relating to years rather in seconds, and it may be quite unsatisfactory to use the wrong time scale. Experience is not always available, in which case a shorter time scale should be tried initially with a gradually lengthening time if no indications are given.

9.4.7 Product sampling (section 8.2)

There are few product lines which can justify 100% sampling and checking; most of these relate to seriousness of failure. In their case the process is long winded, but reliable. In other applications a statistical sample should be taken with the process time being lengthened if no faults are found, or shortened if they are.

9.5 REMOTE COMMUNICATIONS

The transmission of the data – from sensor to analyser – needs to be rapid and reliable. Consider the following possibilities:

- by hand (within a logger, or a sampling bottle for oil samples);
- by direct wire connection;
- by slip ring or FM transmitter from rotating to stationary component;

- by use of existing telecommunication lines (e.g. telephone with a modem);
- by radio, infra-red or other electromagnetic transmission.

The cost of these varies considerably; however, because the initial cost of, say, a radio transmission system may be high does not mean that it is expensive overall. It may be cheaper in the long run if the other techniques will involve the use of a considerable amount of workers' time.

Radio communication is becoming more popular for the larger companies with extensive sites, or where part of the machinery is remote from the main works (e.g. in water operations). Licence-free operation is possible for lower ranges or with the spread spectrum techniques – licence exemption is given by both the UK and the rest of Europe.

Distances covered by the various transmitters–receivers vary with terrain and buildings to a certain amount. For instance, if the transmitter can, as it were, visually see the receiver (line of sight), then a clear signal is obtained over a greater distance. On average the distances transmitted would be of the order of up to

- 1 km for VHF and
- 10 km to 15 km for UHF

although these would be extendible with repeaters.

Transmission rates vary considerably with the different modules and distances covered, and these would have to be matched to the sensor signal generation. Rates could vary, say, from 1200 bits s^{-1} to over 20 000 bits s^{-1}.

CHAPTER 10

Decision making

Data may have been obtained from any one or more of the many sensors and techniques already described. It may have been stored in a data logger, or on a disc or a CD ROM, or even just written out on a log sheet as described in the previous chapter. However, what happens now? How am I to make a decision? At first sight it may be 'obvious', but a second look may reveal that there is much more information available which could lead to a more accurate conclusion. What analysis can be undertaken?

Basically the idea of the analysis is to glean some idea of the progression of the condition of a machine, and with it, possibly, an indication of the likely machine life remaining.

Three general techniques of decision approach are discussed here:

- decisions based on a trend,
- decisions based on an absolute value being exceeded and
- a learning process able to refine the accuracy of the decision.

10.1 TREND ANALYSIS

10.1.1 Basic concept (Fig. 10.1)

Trend analysis is the detection of a change in sensor level with time, which indicates a significant difference from previous levels.

10.1.2 Brief description

A machine trend analysis should examine and smooth the sensor values to determine a normal rate of change. In theory, for a good machine, there might be no change at all, indicating no wear of the machine. However, what is more likely in practice is that very steady, and acceptable, gradual wear is occurring, following the running-in. This is a trend. It is an acceptable trend.

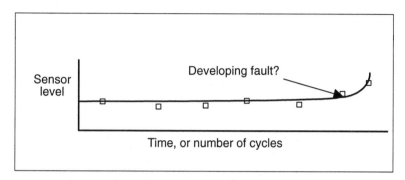

Figure 10.1 Development of a trend.

However, careful monitoring of the trend may show a sudden significant change, indicating the first signs of abnormality, i.e. that associated with poor health developing.

10.1.3 Practical application

It is suitable for all applications where continuous or regular monitoring is present, in particular where the levels of some steady state parameter are being recorded. Examples would be

- temperature of a bearing,
- dimension of a machined product,
- level of fluid in a reservoir and
- content of iron (parts per million) in an oil sample.

10.1.4 Detailed discussion

A rise, or drop, in a value does not necessarily indicate a fault. Indeed, one would expect a gradual change in early life due to running-in. The 'bath-tub' curve (Fig. 10.2) is well known for describing the wear debris generation from bearings – with a high initial value followed by a reasonably rapid decline, followed by a very gradual rise in wear until a more rapid rise indicates the serious break-up of the bearing surfaces.

Care must be taken not to respond too rapidly to a rise in level. For instance if there is quite considerable scatter between sensor readings, then what may be construed as a change for the worse may, in fact, only be a somewhat higher value within the scatter. A smoothing function must be adopted (as shown in Fig. 10.1).

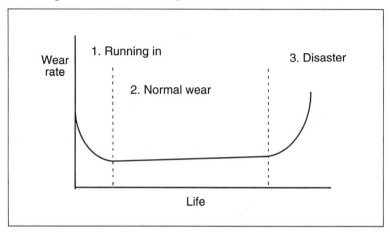

Figure 10.2 The 'bath-tub' wear curve.

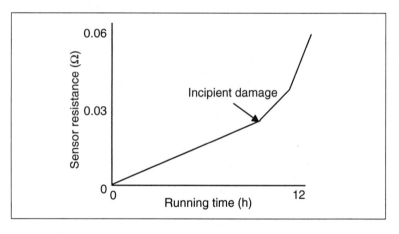

Figure 10.3 Change in sensor resistance (Fulmer).

The point of interest to the person monitoring a machine is that sudden sharper increase in wear rate – and that it has occurred with a consistency beyond any likely scatter in sensor value. This is the stage for action. Figure 10.3 is an example of wear debris detected by an IPH Fulmer monitor, showing the sudden commencement of break-up on a rolling bearing under test.

From the above example it will be apparent that just noticing the change in the rate of wear is not necessarily the important point where incipient damage has irreversibly occurred. In the example there are two points where wear rate has seen an upturn. In other examples there may be several such points and depending on the scale of the vertical axis (y axis) the points of change may or may not seem significant. It is, therefore, essential to evaluate the importance of each change in rate from practical experience or tests.

The determination of the importance can be quite quick if there has been a very steady level previously (either a slope or horizontal) and the new slope starts becoming steeper and steeper in a regular manner. In other words, the rate of change of the slope is consistent. What is happening in this scenario is that the rate of the deterioration of the machine has reached a stage where it is beginning to become progressively worse and worse.

Another means of detecting the seriousness of a trend is by using a second type of sensor. For instance if the main monitor is a vibration sensor then it would be expedient also to attach a wear debris analyser or sampler during initial trials. In this way confirmation of the importance of any changes should be readily obtainable. (Later, the single monitor can be used on its own.) A further bonus of such a test is that of determining which of the two sensor types actually provides the clearest and earliest indication of machine condition. Very often one type has been used historically with little thought of other possibilities.

10.1.5 How to start

Basically, apart from the sensors already chosen, the only need is an efficient means of signal storage (data logging) and a simple software programme. PROMASS (Unipro Ltd), for example, is capable of analysing plant data and of drawing operator attention to significant trends which might indicate equipment deterioration or failure.

10.2 ABSOLUTE STANDARDS

10.2.1 Basic concept (Fig. 10.4)

This is the detection of the sensor level beyond a fixed acceptability level.

10.2.2 Brief description

Two signal levels may be determined; one where the sensor indicates the possibility of malfunction developing, and a more extreme one where the likelihood of a fault is probable. Such levels may be recognized standards or levels appropriate to the situation in hand. The level need not just be higher; in some applications the acceptability of a sensor output will be between a higher and a lower limit. In manufacturing applications the term 'control chart' would be more appropriate.

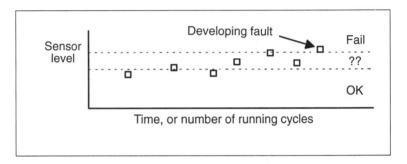

Figure 10.4 Sensor output indicating a fault.

10.2.3 Practical applications

Applications are similar to trend analysis, but are only really suitable where it is known at what level serious conditions are present. Examples of the various kinds would be

- fluid level in a supply reservoir – dropped almost to outlet port, or about to exceed the overflow level,
- temperature of a touchable surface – close to 'burn' level or liable to cause frost bite, and
- concentration of iron in an oil sample – beyond acceptable proportion to other metal content or drops because of high concentration of another wear metal.

Another type of 'absolute' is the visual one. Wear atlases are available which illustrate a range of examples of debris which can be generated from within a machine. Should one such particle be observed on a microscope membrane, then immediately the warning of impending disaster is signalled.

10.2.4 Detailed discussion

The choice of which standard level to use is notoriously difficult. Each machine type is different. However, where the parameters have a direct bearing on the properties of the application (such as temperature or the quality of oil) then a realistic maximum value can be stated for each sensor.

For a total machine operation different components can be monitored and 'stop' signalled should any one exceed the limit – such as with the vibration spectrum in Fig. 10.5 where only one of the frequencies shows a level above the acceptable value.

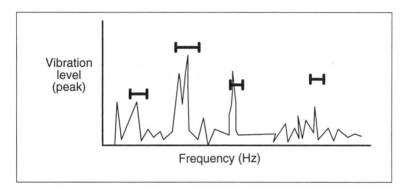

Figure 10.5 Maximum permissible levels at certain frequencies.

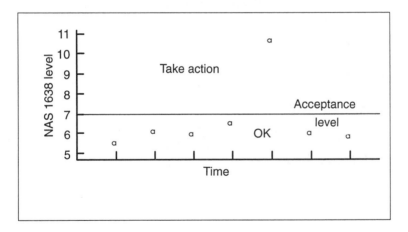

Figure 10.6 Mean debris levels – NAS 7 exceeded (Rover).

There are not many absolute standards available for condition monitoring. Most applications must develop their own critical levels from actual experience. The oil analysis programmes often use parts per million for elemental content limits, but for total debris content the cleanliness standards associated with fluid power applications are now beginning to be used.

One application of the NAS Cleanliness level standard (NAS 1638) is shown in Fig. 10.6 for a robotics condition monitoring check. In this case only one sensor was used, and an absolute level of NAS 7 was taken as a first guess (and later verified as acceptable). On the occasion shown in Fig 10.6 that level was considerably exceeded and the robot was stopped at a convenient time and corrective action taken.

10.2.5 How to start

As with trend monitoring, the need is only a good data logger and a simple software program. In this case, however, it will be necessary to have available the absolute levels or standards which have been agreed.

10.3 NEURAL NETWORKS

10.3.1 Basic concept

This is a system of analysis, based on the human brain, which progressively modifies a variety of different input data, each to a certain amount, in order to make the best assessment of a situation.

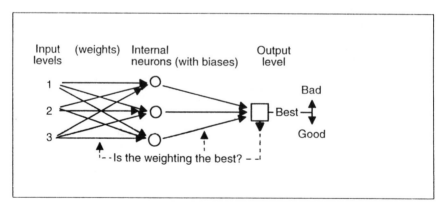

Figure 10.7 A typical neural network system.

10.3.2 Brief description

An artificial neural network (ANN) consists of an input (from one or more sensors) and an output (of one or more lines, although one is the most common for decision making). Between the input and output there are layers of hidden neurons which receive the data, suitably modified, from all the previous layer neurons, and then pass on the levels to the next layer or output. The weighting and biasing are adjusted in an iterative process (over many iterations) until the output is at an acceptable level.

10.3.3 Practical applications

Examples could be

- pattern recognition,
- colour recognition,
- signal processing,
- motor bearing wear,
- internal corrosion and
- pump seal wear.

One application under evaluation is that of the pumping of food products; here the monitoring of acoustic sounds is able, with neural networks, to detect changes in food intensity, cavitation and possibly pump condition (such as impeller imbalance and bearing defects).

10.3.4 Detailed discussion

Before looking at some practical workings of artificial neural networks, it is worth appreciating how valuable they can be from an examination of the human brain. The brain is designed to work as an exceptionally complex neural network and we are merely copying, at a very simple level, what has already been proved so successful. In the case of the human, our five physical senses are constantly being monitored by the brain – but the monitoring is not a simple level assessment but one which is loaded and interacted with by the network. There is considerable feedback and memory involved, with a series of neuron layers and interactive and weighted paths.

The brain decides from

- the intensity, direction and harmonics of sound (detected by the ears), with
- the intensity, position and temperature of feel (detected by touch sensors), with
- the intensity, movement, colour and content of visuals (detected by the eyes), with
- the intensity and substance of gas (detected by the nose) with
- the intensity and substance of ingested mass (detected by the taste buds)

what situation has developed and what action is necessary. Also, the process speeds up and is the more effective the greater the number of similar (but not necessarily the same) previous observations. In other words, the brain has learnt over the life of the subject the levels of weighting and bias to give to each aspect of the detection. It also is constantly updating and modifying the weighting and biasing from each new experience.

In practice an artificial neural network comprises a computer network with

- any number of inputs (input neuron layers connected to transducers),
- internal neuron layers (hidden neuron layers not directly connected to the outside),
- the ability to weight the transference of signals between neurons and to bias the importance of each neuron in each layer and
- any number of outputs (output neuron layers connected to signals or controllers).

The ANN, in the learning mode, endeavours to set itself up with the available series of inputs and the required series of outputs, with the minimum number of hidden neuron layers (the smaller the number of layers the faster the analysis). Figure 10.7 shows a simple ANN with one hidden layer and a single output.

Initial work is undertaken to decide the right loading necessary to obtain the most accurate result. Having decided the best weighting and bias within the available network, the network can continue to be used to determine accurately machine condition – 'bad' or 'good' – based on the sensors' inputs. Both analogue and digital inputs can be interrelated. There is a process of improvement where the weighting and biases are changed if a better result can be achieved thereby.

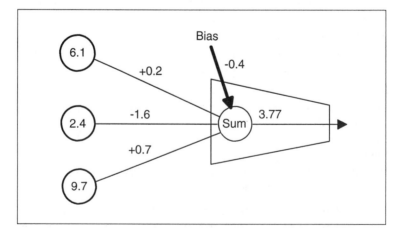

Figure 10.8 Typical neuron set of values.

If we look at Fig. 10.7 and take the top hidden neuron which has three inputs which can be weighted in different ways, and the whole neuron result can be biased, then a practical example is shown in Fig. 10.8. Here is a typical numerical set of values for an initial test. The output from the neuron is calculated from a summation of each of the input neuron values times the weighting, all shifted by the bias, i.e.

$$6.1 \times 0.2 - 2.4 \times 1.6 + 9.7 \times 0.7 - 0.4 = 3.77.$$

There are several means of arranging the neuron layers, and no doubt others will emerge as more research is undertaken.

Common methods in 1995 have included the following:

- **Multilayer perceptrons** (MLPs)
  As the name suggests this neural network has more than one hidden layer. This is the standard method of neural networks. It is 'supervised' in the sense of being trained with one or more examples of the correct output from a sample set of input; thus it learns the best weighting automatically.
- **Radial basis function (RBF) network**
  This is virtually a single hidden layer MLP (as in Fig. 10.7 but without the weighting between the input and hidden layers); again it is supervised. However, although it has a speedier optimization of the weighting between the hidden and output layers, it is slower at deciding an optimum vector position for the hidden neuron. This means the time of 'learning' can be just as long as with the MLP network, but it could be better when the input data come in clusters. It is particularly valuable for those cases where the output data represent discrete categories such as with fault diagnosis.
- **Kohonen self-organising feature maps**
  This is somewhat different to the MLP and RBF networks. It has 'unsupervised' learning in the sense that only input data are presented, but, and this is the key point, from remembering previous input data it can determine whether changes are significant. The algorithm discovers patterns and associations in data when outwardly there is no apparent connection. This is particularly valuable for condition monitoring where a machine goes from an acceptable condition to one which is possibly unacceptable. It can be used for spectral analysis as well as more general data, detecting those unusual occurrences and providing the necessary signal or alarm.
- **Rho network**
  This network determines the probability that the input pattern belongs to a specific pattern class, and hence it can detect when a

process moves from being acceptable (usual) to being unacceptable (unusual). Users can have parallel rho nets to generate on-line probabilities for a complete set of output categories.

From the above it may be obvious to us, that whilst the MLP and RBF networks can give hard and fast rules, the Kohonen and rho can only sense that a difference has occurred. On the other hand, there are many condition monitoring situations where it is just not possible, or it is too costly, to obtain output data (called 'target' data) for a fault condition.

The user needs to contact suppliers of neural software to obtain the latest and most relevant for the particular application in mind. The Condition Monitoring & Sound Assessment Club (based at AEA Technology, Harwell) would also be able to give the latest information. It is a DTI-sponsored group of companies and has looked at helicopter gear boxes, gas turbines and flight data processing amongst other applications. DTI help can be reached on the Internet, called the Neuro Computing Web service, on page http://www.globalweb.co.uk/nctt/. and includes information on software and document archives, on-line discussion groups, applications demonstrator clubs, regional activity and technology transfers.

Not all monitoring applications require neural networks. They certainly do help, however, where

- the signals are indistinct (low level),
- the signals are damped (muffled),
- there is a considerable amount of 'noise' from other components and machinery or
- the signals generated are so complex that current analysis techniques are unable to fully cope.

In the case of a machine, with several sensors in different positions and of different kind, the ANN analysis starts with an initial calibration with some known examples. This is the basic setting up of the network based on past experience. This is the 'supervised' training. Thereafter it should be possible to use a direct feedback algorithm with 'unsupervised' training providing the necessary updating and improvement.

Where no experience has been logged from past work, then the unsupervised training starts immediately with a random selection of weightings.

There are a number of related topics to an ANN. ANN analysis is a development of artificial intelligence (AI), expert systems and, in general, knowledge-based systems (KBSs). It is sometimes referred to as an adaptive network system (ANS) analysis.

Fuzzy logic is another allied idea which deals with systems where the true–false logic is inadequate. It allows the process engineer to describe a system in simple linguistic terms (such as 'fairly low' or 'rather fast') and then to see a hardening up of that imprecise input data. In other words, fuzzy logic is an uncomplicated means of adding intelligence to a wide variety of products to improve features and to increase efficiency. The performance of this analysis is done by means of a set of fuzzy rules which are processed by a precise mathematical algorithm. In practice, any one of these is provided as a microcontroller ROM or EPROM. As far as the user is concerned he or she merely uses fuzzy logic vocabulary and concepts to describe the application. Fuzzy logic can be developed with neural networks; the data structures are automatically translated into a proprietary neural network representation on which training algorithms can be applied to tailor the fuzzy knowledge base to specific applications. The expression 'neuro-fuzzy' is sometimes used.

It has been suggested that the combination of ANN and fuzzy logic has considerable possibilities with high financial returns. For instance if a rho network is trained on a 'good' and a 'bad' scenario it can determine whether a process problem exists, but it cannot necessarily, on its own, classify the root cause of the problem. To diagnose the problem fully the rho network probability output can be 'fuzzified' so that linguistic comments can be combined with other evidences of process abnormalities.

10.3.5 How to start

Besides the sensors, an ANN software package is essential. There are a number of these available but they differ considerably in complexity. For instance, the capacity of the PC used is likely to be a major consideration owing to the extensive computing power required; even a Pentium may not be sufficient in the more advanced cases.

Suppose, now, that a machine needs to be monitored and the ANN is going to help in developing the monitor, which, say, is based on vibration detection. It will be necessary first to obtain two sets of data, preferably extremes – one being totally 'acceptable' and the other a complete 'disaster'. These constitute the initial training set. The set can then be used to teach a neural network (of the supervised type) to recognize two levels of decision – 'pass' and 'fail'. Without this training set it is more difficult to obtain any sensible result until a lot of experience has built up (with the 'unsupervised' learning).

That is the simple start. However, there is likely to be more than one failure mode, possibly sensed by more than one sensor, and these will have to be included if a successful decision on the quality of the total machine is to be made.

Vibration is not an easy example to take. It must be remembered that the input to the neuron is a number, and hence a vibrating signal must be reduced to a single number by suitable pre-processing. Such would be the RMS value or the level of the fifth harmonic after an FFT. Steady-state monitoring is likely to be much easier but not as rewarding.

In addition to pre-processing, the total neural network analysis needs to determine the optimum number of layers and neurons (hidden as well as outside) to be able to predict the result. Examples in the literature (Pei *et al.*, 1995; Er *et al.*, 1995) are

- 12–15–1 (i.e. layers of input, hidden and output neurons) for the prediction of the vibration peak-to-peak levels on a 200 MW turbo-generator, and
- 53–10–1 for the on-line testing of loud-speakers, which took 40 min for a 486DX 50 PC with 100 patterns.

Another valuable use for ANNs is the monitoring of monitoring sensors. By connecting several sensors on a machine to a single network, the likelihood of error can be detected very rapidly, such as if one of the sensors is going off calibration. The sensors need not be the same type, but all need to relate to machine acceptability.

PART THREE

Experiences in Monitoring

CHAPTER 11

A typical CME analysis

This chapter serves as an example of how to deal with a specific application. It could be one's own situation which needs the installation of a monitor, or someone else's where advice has been sought. Perhaps we have been approached to advise where a problem has already occurred, or it could be a far-sighted manager who has insisted that condition monitoring must be part of the overall process.

How do we go about installing a suitable monitoring system?

This is where we put into practice what we have learnt in the previous ten chapters. In this chapter the overall scheme will be approached according to the CME technique introduced in Chapter 2. However, let us introduce the problem first.

We could take as an example a complete plant process, but that would be unnecessarily complex and would probably lose us in the details. We can understand the idea much better by looking at a single unit which in itself is complex but easier to take in – a high-speed, high-pressure hydraulic pump. (In passing it should be mentioned that although this example is based on a real pump in a real situation, in no way was the pump considered faulty in itself. The reasons for the failures mentioned were entirely due to misuse of the pump by inexperienced personnel.)

11.1 C – COMPONENTS WHICH FAIL

The pump is illustrated in Fig. 11.1. While the overall pump performance will be considered in the next section, it is also important to review what parts within the pump could possibly fail. The cut-away view suggests the 14 different components described in Table 11.1. Note the kind of failures which are considered – all sorts.

With such a large number of potential faults we may well wonder whether the pump will ever work. The comments relate to what had already been experienced before the suggestion that monitoring be

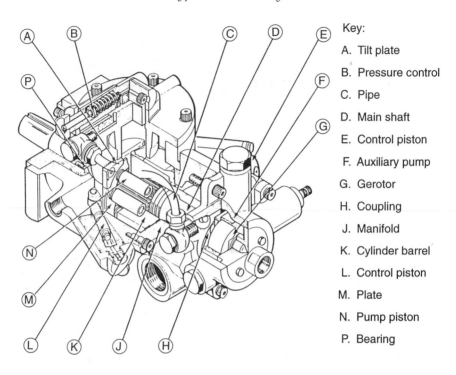

Key:

A. Tilt plate

B. Pressure control

C. Pipe

D. Main shaft

E. Control piston

F. Auxiliary pump

G. Gerotor

H. Coupling

J. Manifold

K. Cylinder barrel

L. Control piston

M. Plate

N. Pump piston

P. Bearing

Figure 11.1 The pump for monitoring.

applied. They help to give a measure of likelihood and relative importance to each component. (In a plant situation there will be some
components which would not cause a serious problem if they did fail
and the probability of such an occurrence may well be extremely remote.
In that case one would reject them from the list at this early stage.)

Some 'failures' are slow, while some are sudden and catastrophic.
Some are primary and some are secondary (i.e. they occur only because
a primary failure has already occurred). We need to put into an order of
priority those components which we feel should (or, taking all consider-
ations into account, might) be monitored. The list in Table 11.2 is one set
of suggestions, covering the six most important.

11.2 M – MONITORING METHODS WHICH COULD DETECT FAULTS

Now we need to put into practice the knowledge we have gained from
Part Two of this book. What are the monitoring types which might be
appropriate?

Table 11.1 Components which may possibly fail

Component	Purpose	Failure modes	Comments
(A) Tilt plate	Rotated by control pistons	Wear and jamming of sliding surfaces	Yes, jammed owing to wrong oil
(B) Pressure compensator control	Setting for output pressure	Wear and excessive leakage	Yes, seriously worn by contaminated oil
(C) Pipe	Oil cooling supply line	Fracture	Unlikely if correct support
(D) Main shaft	Central drive	Wear	Very gradual
(E) Control piston (1 of 2)	Actioned by compensator	Wear and seizure	Some wear from contaminated oil
(F) Auxiliary pump	Coupled to main shaft	Wear and leakage	Possibly OK
(G) Gerotor (gear with plain bearing)	Provides low pressure oil flow to the casing (coolant)	Wear and bearing break-up	Gradual wear only
(H) Coupling	Drive connection	Fracture	Secondary fault
(J) Manifold (plain bearing)	Includes the inlet–outlet ports	Cavitation, erosion and break up of bearing	Yes, seriously destroyed by inlet problems
(K) Cylinder barrel	For the nine pistons	Cavitation, erosion and wear	OK if correctly dimensioned
(L) Control piston (1 of 2)	Actioned by compensator	Wear and seizure	Some wear from contaminated oil
(M) Piston retraction plate	Holds pistons to tilt plate	Breakage	Probably OK
(N) Main pistons (9)	Elastohydrodynamic lubrication of slipper pads	Wear, scoring, blockage	Contaminant effect
(P) Roller Bearing	Tapered to take end thrust	Wear, spalling and break-up	Lubrication fault

Table 11.2 Order of priority of component monitoring

Component	Reason	Component	Reason
1. Main pistons	A fault would seriously reduce the output	4. Roller bearing	If this should totally fail so would the pump
2. Pressure control piston	Wear of this control would cause secondary faults	5. Tilt plate	Should this jam the whole concept of pressure control is lost
3. Manifold block	Severe cavitation means loss of power	6. Plain bearings	Overheating and secondary damage would occur

There are the types associated with what enters the pump (Chapter 6), with what is happening within the pump (Chapter 7) and with what is coming out of the pump (Chapter 8). Taking the six components mentioned in Table 11.2 we can give an initial likelihood score to the various possibilities on the chart taken from Appendix A (Table A.2). This is shown in Table 11.3.

Table 11.3 Check list for possible monitoring methods

Type of monitor	Components considered for monitoring						Monitor preference and comments	
	1	2	3	4	5	6		
Environment:								
Temperature								
Relative humidity								
Pressure								
Light								
Vibration								
Air velocity								
Pollution								
Radiation								
Gas								
Power	7		6		6		If repeatable service	
Other (input)								
Human:								
Colour								
Smell								
Vibration	4	6	7	6			If signal clear enough	
Ultrasonics	5	7	8	7		4	If signal clear enough	
Wear debris	8	8	7	6		7	If before filters	✓
Oil					4		Limited	
Thermography				7			Difficult to get in	
Leakage	7	4	7				Internal?	
Corrosion			7				Difficult to get in	
Steady state:								
Pressure	7	8	4				May not be adequate	✓
Flow	7				7		May not be adequate	
Level								
Speed				4	5	5	If repeatable service	
Temperature				4		6	Difficult to get close	
Other (within)								
Performance	5	8	8		8		If repeatable service	✓
Product variants								
Other (output)								

Possibility rating: 1, not very likely; 2, just a possibility; 3, a possibility; 4, 50:50; 5, more than just likely; 6, likely; 7, very likely; 8, high likely indeed.

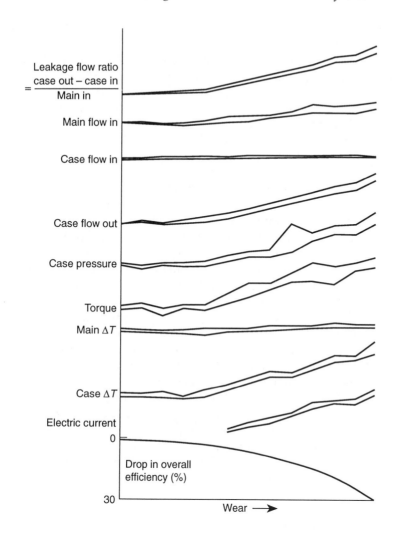

Figure 11.2 Relative significance of the monitoring methods on the pump.

The list is still quite large but a selection of preferred monitors is beginning to appear. What we still do not know is how well each will perform in the actual situation. Where the monitor is complex and expensive (e.g. the vibration monitor), then what should be done is an on-the-job example run. Most suppliers of expensive monitoring equipment will be happy to do this. (An alternative is to hire the monitor from a hire company, several of which now have a specific condition monitoring section.)

Cheaper monitors must be chosen on their specification and personal recommendation of existing users. There are some monitors on the market which have sold extremely well owing to exceptionally good publicity and advertising, but in practice many have found their way to the 'pending' tray or a cupboard for discarded gear. Check with someone who uses the cheaper monitor before committing the company not only to its purchase but also to its long-term reliance on it as a warning system. If the warnings do not come at the right time the costs of a misplaced faith may well be disastrous.

For the example in hand, the pump, tests were undertaken with vibration, ultrasonics, wear debris, temperature, power and performance. From those tests it was apparent that the vibration and ultrasonics would require very expensive monitoring equipment due to the proximity of other systems. It could be done, but it would be expensive.

Temperature testing on the bearing housing was again somewhat obscure owing to the thickness of the housing which caused heat dispersion in an unclear manner.

Power, performance, temperature and wear debris all conveyed significant changes when severe contamination was added to the fluid. This is shown in Fig. 11.2 with an indication of the drop in pump efficiency at each stage, due to the wear which has taken place. (Note that wear debris was not monitored continuously because of the type of artificial test which was undertaken.) Although all except 'case flow in' showed a change with deterioration of the pump, it was the ability to detect the gradual and consistent change in monitoring parameter which was to decide the optimum (next section). Some monitoring detected wide variations and hence the graph limits are wider in some places.

11.3 E – ECONOMICS OF COSTS INVOLVED

Costs have already been mentioned in the previous section, as regards the monitors. However, a more detailed analysis is necessary, as mentioned in Chapter 2 and summarized here:

- cost of item being monitored;
- cost of downtime should the item fail;
- cost of monitor (initial purchase);
- cost of monitoring (consumables and personnel in maintenance and analysis).

In effect what we need to do is to balance the cost of a loss (without a monitor) with the cost incurred in fitting and using a monitor. Is it worthwhile? This is the terotechnological concept introduced in Chapter 5: the consideration of total cost rather than today's cash flow.

Consider now the pump situation in question (the figures are an approximation only):

- cost of pump –   £1,000;
- cost of downtime –   a complete ship and all its crew
                                   (at least in a state of war);
- cost of monitor –

  1. vibration or complex ultrasonics, £50,000,
  2. performance, £2,000 + cost of software,
  3. power, £2,000,
  4. wear debris, £8,000;

- cost of monitoring –

  1. vibration or complex ultrasonics (small if on-line to a PC),
  2. performance (small if adequate software),
  3. power (small),
  4. wear debris (medium to small depending on type).

There are also the reliability and effectiveness of the monitor to take into consideration – a cheap monitor may not be adequate.

In fact, the ship operator chose first the performance option in that it was easy to fit to existing installations and was reasonable in cost both initially and long term. The type of performance chosen related to the internal leakage detected by flowmeters on the inlet and output side of the pump (the leakage flow ratio in Fig. 11.2). In order to detect also bearing break-up, the second choice (in addition) would be a wear debris monitor.

CHAPTER 12

Some pitfalls

This chapter is really a warning.

Condition monitoring may be common sense, but not all sense is common. In other words, if we are to have any chance of making a sensible choice in our monitoring, we need to know

- what is available,
- what can be done,
- what cannot be done and
- how successful or otherwise the monitoring has been.

Five sections will look at various reasons why condition monitoring has failed through lack of understanding.

12.1 INADEQUATE TRAINING IN MONITORING AND MAINTENANCE

This is what this book is all about – making the whole subject of condition monitoring known and understood: not just parts of it, but all of it, or at least, glimpses of all of it; not just as individual ideas but how the whole field fits together. There is sufficient information here to start and to avoid being 'slightly misled' by the over-enthusiastic salesperson who has but one product to sell.

However, good though this book may be, a book can never give the clarity of understanding that an individual gains through field experience. There are many courses which are available, from half-day seminars to 12-month M.Sc. degrees to give the in-depth experiential training (with the book). Quite often the course has been designed to help the uninitiated to be able to sense what monitoring system is best for the machinery in question. It is worth considering whether one or more personnel from the company attend these. The alternative is to get a training group or consultant to arrange an in-house seminar.

Some of the courses are specific, in the sense of being on corrosion or on wear debris or on vibration or on thermography, etc. They are valuable in that they bring up to date the subject, with the very latest in the field. There are other courses which are on maintenance appropriate to one's own industry and including condition monitoring as a major part of the training. There are also courses which are devoted solely to the subject of condition monitoring in its entirety. A caution should be stated here as regards the 'condition monitoring' courses; not all are as complete as might be suggested by the title. The course syllabus should be checked to see that they do cover all the subjects mentioned in this book; it is not uncommon for an establishment to have an emphasis (often connected to the field of the professor) and the other fields are given much less prominence.

The following are some typical examples of syllabuses in maintenance and monitoring which may give ideas of what to attend, or even what to organise in one's own establishment. Note the difference between a single type of monitoring and a more general course. There is also the possibility of attending part of a larger course in order to obtain greater depth in that subject alone.

(a) IRD Mechanalysis

- Three-day vibration technology course VT1 (1995):

  1. benefits and use of predictive maintenance;
  2. vibration technology;
  3. definition of vibration and measurement parameters;
  4. how much vibration is too much vibration;
  5. introduction to swept filter and FFT analysis;
  6. how to obtain good vibration data;
  7. phase measurement and analysis;
  8. how to diagnose machinery vibration;
  9. introduction to balancing.

- Three-day vibration technology course VT2 (1995):

  1. understanding all the factors which combine to determine the vibration characteristics of a machine;
  2. how vibration analysis instrumentation, signal processing and measurement techniques can affect analysis results;
  3. special analysis techniques to solve unusual vibration problems;
  4. tests to distinguish between machine faults having similar vibration characteristics;
  5. analysis of data from actual case histories and the procedures applied to diagnose machinery problems.

- Two-day data collector and software course DC1 (1995):

  1. introduction to basic DOS commands;
  2. applications of the techniques learned on the VT1 course to set correct measurement parameters and alarm levels;
  3. planning and setting up a measurement route efficiently;
  4. the use and interpretation of computerized reports;
  5. methods of automating the software by utilizing batch files for set-up, configuration, operation and report generation.

(b) Longlands College of Further Education

- Five-day short course 'Introduction to Condition Monitoring' (1990):

  1. overview of techniques and their applications;
  2. demonstration of vibration analysis;
  3. hands-on practical data collection methods;
  4. demonstration of wear debris analysis;
  5. hands-on practical work in wear debris analysis;
  6. demonstration of oil condition analysis;
  7. practical work on oil condition analysis;
  8. demonstration of electric motor monitor;
  9. data collection by motor monitor;
  10. thermographic imaging theory and practice;
  11. review of techniques.

- Three-day short course 'Designing a Condition Monitoring System' (1990):

  1. introduction – overview of techniques and the application of plant condition monitoring;
  2. development of reliable low cost maintenance systems;
  3. application of systems approach to condition monitoring;
  4. economics of condition monitoring;
  5. fundamental aspects of a system;
  6. equipment used with a system;
  7. examples of a condition monitoring system;
  8. future developments.

- Ten-day short course 'Comprehensive Condition Monitoring' (1990):

  1. vibration;
  2. thermography;
  3. wear debris;
  4. acoustic ranging;
  5. acoustic emission;

6. acoustics and noise;
7. balancing;
8. oil condition analysis;
9. electric motor monitor;
10. vibration analysis–diagnosis.

(c) Nikat Associates, Chester

- Various two- or three-day courses on total productive maintenance techniques, predictive maintenance techniques, machinery diagnostics, advanced machinery diagnostics, machinery shaft alignment and wear debris and oil particle analysis.

(d) Southampton Institute of Higher Education

- M.Sc. in condition monitoring (1994–95):
  1. principles and applications of condition monitoring (philosophy and justification (5%), instrumentation technology (30%), vibration and noise monitoring (25%), wear debris monitoring (10%), corrosion monitoring (10%), thermal and force monitoring (20%));
  2. principles and applications of integrated maintenance management (maintenance planning (20%), reliability centred maintenance (45%), maintenance organisation (20%), applications (15%));
  3. principles and applications of diagnostic technology (cause and effect of engineering failures (10%), failure detection and monitoring techniques (40%), metallurgical failure modes (35%), non-destructive testing techniques (15%));
  4. technology management (business environment (40%), resource management (40%), case study (20%));
  5. project management and project preparation (project management (60%), sources of information (10%), project (10%), data collection (10%), reporting (10%));
  6. project.

(e) University College of Swansea

- Five 20-hour modules (part of the Health Monitoring options in the M.Sc.). These are available for non-university members to attend. In addition there are a further 100 hours total for practical examples, and industrial visits (1995):
  1. rotor dynamics;
  2. machinery condition monitoring;
  3. oil–wear debris;

4. noise–vibration;
5. reliability.

(f) Wolfson Maintenance

- Three-day short course 'Cost Effective Maintenance' (1993):

  1. maintenance – management economics, including what is maintenance?, terotechnology, plant availability and maintenance strategies available;
  2. the way forward, including condition-based maintenance, organisation, quality management and computers in maintenance;
  3. maintenance plan, including structured maintenance, preparation of the maintenance plan, monitoring the effectiveness and allocation of resources.

- Four-day short course 'Condition-Based Maintenance' (1993):

  1. the management of condition-based maintenance (25%);
  2. inspection, process, thermal and optical monitoring techniques (25%);
  3. wear, erosion and corrosion monitoring (25%);
  4. vibration monitoring (25%).

12.2 INADEQUATE UNDERSTANDING OF THE PLANT AND MACHINERY

Having been trained in, or become aware of, the subject of condition monitoring, it can only be successfully applied if the plant and machinery is fully understood. This is the 'C' of the 'CME', the 'components which fail' part. We need to know the machine.

What sort of mess do we think a doctor would make if he or she knew only about available drugs and nothing about the physiology of the human body? Quite frequently this is precisely what happens with a salesperson who knows the monitor but does not know the machine. The result is a disaster. It is essential for design personnel within a company to be part of the team which examines and defines which monitoring system or systems are to be incorporated (Table 2.1).

The types of features which must be known are those which relate to the purpose of the machinery and those which may be in trouble – discussed on the very first page of the first chapter of this book.

12.3 INADEQUATE COMMUNICATION–SAMPLING

This is connected with the monitoring itself. It has been mentioned already as a key factor, but it needs mentioning again because it is so frequently the reason for poor and ineffective monitoring. The high quality expensive machine is there – crying out for help. The high quality and possibly expensive monitor is there – already to show what has to be done. However, the communication between the two is almost zero.

The link is just as important as the machine and the monitor. Skilled personnel and carefully prepared equipment are both vital in achieving accurate and reliable communication – how the signals of distress are transmitted and how they are recorded.

This subject is covered in Chapter 9 in a variety of ways. However, to emphasize what we are saying, here are two real examples which make the point. Fig. 12.1 is a true picture of various 'super clean' containers which were used to send fluid samples for analysis at the Bath Fluid Power Centre (before advice from the Centre had been sought). Seeing these, we may well doubt whether the sample of oil itself had been properly taken, but even if it had been scrupulously extracted from a machine, to place it in a container probably containing metal and other dirt particles as well as some other fluid which would not mix with the oil completely nullifies the result.

Figure 12.1 Examples of fluid sample containers.

Another failure occurred in a mine example where an efficient portable vibration analyser was used to monitor the vibration of a slow speed high pressure ram pump running at 512 rev min^{-1}. However, the lower frequency limit of the analyser was 10Hz, well above the actual once per cycle signal of 8.5Hz and they were looking for the out-of-balance at once per revolution. The communication was zero.

12.4 INADEQUATE CONTROL

Quite often the supplier of condition monitoring equipment is unable to provide more than the initial setting-up training for the customer. Thereafter, the operator is left on his or her own to use the equipment as well as possible. This is not good enough.

Condition monitoring equipment is not the same as a manufacturing machine – a miller or a gear hobber, a press or a forge, or whatever. The condition monitor is much more flexible and devious, and is expected to vary in its output from time to time. What happens when a thermographic picture appears a little different to what it was yesterday? The immediate response might be a fault, but is the fault in the monitor or the machine, or is it a fault at all? Is the difference to be expected as a result of poor usage? Is it all to be expected in the range of operation of the machine?

The point here is that of being sure that the monitor is being used correctly for whatever application is currently applied. Written instructions are rarely adequate in complex condition monitoring unless the operator is committed to following them, or asking questions if he or she is unclear.

On the subject of oil sampling, there was the case of a keen company maintenance engineer who, in order to save time (and expense), decided instead of taking a sample from the machine every day, to take an extra large quantity on the first day and just divide it up into separate lots for the laboratory to analyse, one per day. It was only spotted by a more knowledgeable engineer wondering why the results were so consistent.

12.5 INADEQUATE RESPONSE

So, the monitor has given a signal which has been verified as significant; what happens now? There have been cases where monitors have been fitted 'reluctantly'. In other words, while the management consider them valuable and will keep the machinery running much longer and with greater reliability and efficiency, the machine operator wants to have the last say as to whether a machine should be stopped or left running. After all, if the wage packet depends on the quantity of the output on that shift, it will be selfishly important to leave any break-down to the next one.

Because of such lack of co-operation and understanding, not only is the condition monitor almost totally useless, but the machinery is seen with a false sense of security by the management.

There must be a means of ensuring that action follows a monitor's signal. It may occur without problem just because there is a good relationship throughout the company. On the other hand, to be totally confident, the intermediaries (or the sampling deficiencies) can be bypassed by transmitting the signal directly to a central control position.

CHAPTER 13

Examples

Examples of the use of various monitors have already been given in many places within the earlier text (the Index gives further information). However, the purpose of this chapter is to highlight a brief selection of actual examples over the range of monitors, in order to encourage the reader to use condition monitoring. There are examples where they did not succeed – for obvious reasons, as it later turned out. Most, however, are examples where, following careful preparation, the monitoring has proved well worthwhile.

Many conference papers have presented innumerable examples of monitors being fitted to machinery. Quite frequently, however, the majority of the paper is devoted to the philosophy behind the monitoring, and while that is important, in this chapter it is not necessary. There are also papers which illustrate, at least to this current author, that the management team has done the right thing and an efficient and correct monitor has been fitted. However, they do not necessarily prove this fact because no failure has been discovered in the short time of use. Time is needed to prove the point.

Most of the examples here are of cases where a failure has been detected. Two things will be of direct interest to the reader:

- the application;
- the monitor used.

A list of ten examples not mentioned elsewhere in the book is given in Table 13.1. They are shown under the above two headings, and given in the order that the subjects were covered in Section Two of this book.

13.1 Combustion gas analysis (Smith, 1992)

Condition and detection

BP Research has been involved in the development of optimum combustion processes. In order not only to determine that condition but also to maintain the machinery at the optimum level, they examine the gas emitted. The system measures the concentration of CO_2, CO, NO_x, SO_x and unburned hydrocarbons. However, all combustion processes produce considerable quantities of water, which can adversely affect analysis of the gas; a heated sample line, therefore, maintains the gas sample above the dewpoint, so preventing condensation.

Table 13.1 Summary of examples discussed in Chapter 13

Number	Application	Monitor
1	Combustion process (gas)	Environment (ADC custom built)
2	Sewage (screw lift pumps)	Power (Fenner Monitor 360)
3	Process plant (fan)	Vibration (IRD Mechanalysis Inc. Amethyst monitor)
4	Water treatment (gear box)	Ultrasonics (Holroyd Instruments)
5	Power plant (diesel engine)	Wear debris (ferrography)
6	IBM microelectronics (pump)	Oil analysis (CSI OilView® 5100)
7	Paper manufacturer (valve)	Thermography (FSI)
8	British Steel (process water)	Leakage (Palmer Environmental MK4 ground microphone)
9	Offshore (pipe lines)	Corrosion (CML electrochemical impedance spectroscopy)
10	Waterworks (pump)	Efficiency (AEMS Yatesmeter)

Result

The sample has successfully been extracted directly from the source and, by using a combined filter and probe, and sample line, the gas is analysed remote from the source. In this way the customized analysers are not subject to excessive temperatures, vibration or attack from acidic gases.

13.2 Screw lift pump (Fenner, 1991)

Condition and detection

Screw lift pumps are used to remove the effluent from woollen mills and other sources, in the border country of Scotland. However, they can easily become partly jammed, or conversely, the effluent jams upstream and cavitation can occur. While the pumping does continue, it does so very inefficiently and becomes excessively expensive. The Fenner Monitor 360 is able to detect overload (or underload) of machinery driven by a three-phase electric motor; it can also monitor pressure (and other sensor parameters), and was hence fitted to monitor this particular application (Fig. 6.5).

Result

The Monitor 360 was attached to the pump control such that should a power excess be reached the pump would be switched off and the stand-by pump started up; at the same time it would signal the first level of failure. This has happened several times, and because of the repeated awareness of one pump, in particular, showing this characteristic the head bearing was examined and replaced before expensive repair work was necessary.

13.3 Process plant clinker cooler fan (Mathur, 1994)

Condition and detection

In normal operation the vibration level (velocity) of the fan bearing was around 2–3 mm s^{-1}. The simple vibration detector device measured an increase to 11.3 mm s^{-1} which was way above the alarm value of 6.4 mm s^{-1}. The Amethyst portable analyser was connected and diagnosed that there was a defect on the inner bearing of such magnitude that the bearing should be replaced at the earliest opportunity.

Result

10 days later was a convenient time. The bearing was removed, found to have a defect on the inner race, replaced and the bearing vibration reduced to the usual and acceptable 2 mm s^{-1}.

13.4 Aerator gear box (Holroyd and Randall, 1994)

Condition and detection

Initially a planned maintenance arrangement had been formulated to refurbish 8 gear boxes per year out of the total 80 on site – i.e. a 10 year cycle. However, in order to maximize the use of the gear boxes and to prevent untimely failures an ultrasonic Machine Health Checker was employed which could monitor all 80 gear boxes within 2 h, once a month.

Result

The results in general followed two patterns, one which varied slightly about a mean value (Fig. 13.1, condition a) which related to a totally acceptable gear box, and the other which showed a gradual upward trend of deterioration (Fig. 13.1, condition b). However, one gear box

(Fig. 13.1, condition c) gave an excessively rapid rise in ultrasonic signal which in five months exceeded 40 dB above the normal. Unfortunately the engineer in charge could not feel any change in temperature, and the vibration level and noise were not obviously increasing, and so he allowed the box to continue to run. Two weeks later the gear box seized and was then found to be a total write-off.

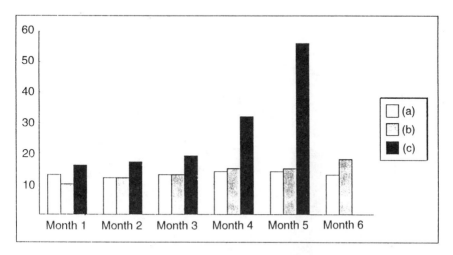

Figure 13.1 Ultrasonic levels (decibels) over a 6 month period for three machine conditions (a–c).

13.5 Diesel engine wear (Enwald and Liu, 1994)

Conditions and detection

Environmental conditions for the power plants in the Guangdong Province are around 90% relative humidity and 40°C ambient temperature. With heavy loads and 20 h per day running, it is not surprising that severe wear can occur on the diesel engines. Most prone are the crankshaft bearings, cylinder walls, piston rings and the camshafts (lobes and gears). It was decided to monitor the situation by means of wear debris analysis of sampled oil using ferrography and bichromatic microscope. This was later re-analysed using an expert system.

Result

The 40 mm position on the slide was taken as the comparative region looking at particles larger than 2 μm. A count was made of the particles within the area of view and the particles were classified in terms of wear type (i.e. fatigue, cutting, adhesive and abrasive) and appearance (i.e. copper, cast iron–steel, black oxides, red oxides and sand and dust).

Both the human analysis and the automatic image analysis were able to determine excessive levels of wear debris according to the different regions of the diesel system, as shown in Fig. 13.2.

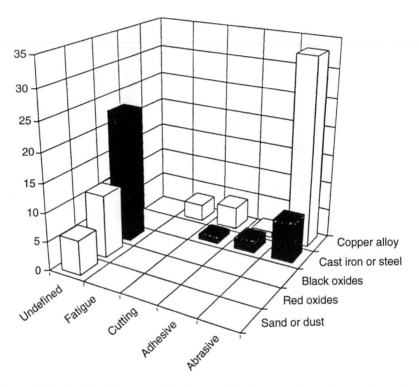

Figure 13.2 Distribution of particles detected by ferrography on diesel oil samples (Enwald and Liu, 1994).

13.6 IBM vacuum pumps (Garvey, 1994)

Condition and detection

An Edwards single-stage rotary vane mechanical vacuum pump operates for 5–15 min at a time, 10 times a day, to evacuate a high vacuum chamber used in the metal evaporation processes. Oil samples were analysed by means of an OilView® in-shop oil analyser at the end of each run.

Result

The initial levels recorded and those after one 15 min run are shown in Table 13.2. The dramatic increase initiated an immediate repair of the pump at a cost of $1,000. Left undetected the pump would have failed and for a replacement the cost would have been $16,000. The damage was due to coolant contamination.

Table 13.2 Oil sample analysis of Edwards pump

Sample date	Oil life index	Corrosion index	Contaminant index	Ferrous index	Large contaminant	Viscosity
Initial	0	0	0	0	None	NA
12 January 93	759.3	34.3	Critical	Critical	Ferrous and non-ferrous	NA

13.7 Steam valve in paper processing (Data Cell, 1994)

Condition and detection

A paper manufacturer had had problems with moisture streaking on the finished product. The process is a long one, but because thermography can sense very small differences in temperature, it was suggested that the process could be checked throughout with thermographic images.

Result

In a half-day survey it was discovered that the streaking, seen in Fig. 13.3 as temperature lines, first occurred part way through the high PLI end of the process. Further investigation revealed that the fault was a partly clogged steam valve. For the cost of 4 h with a thermographic camera and operator, and the immediate correction to the system, many tons of otherwise rejected production had been saved.

13.8 Sunken water pipe leakage (Anonymous, 1995)

Condition and detection

British Steel's Scunthorpe works covers almost 700 ha, producing up to 10 000 t of liquid steel per day; but for each tonne of steel produced, it

Figure 13.3 Moisture streaking of paper seen by thermography (FS1–Data Cell).

has also been losing 2 t of water. Extensive recycling is used, and some losses are expected from the process, but this enormous loss prompted the need for leakage detection below ground.

Detection previously was done by expensive digging around, but this time the Palmer variable filter selection MK4 microphone was used. This could be arranged to filter out the background noise and highlight only the leakage sounds.

Result

It was understood that the pipes under question were steel and hence only the high frequencies needed to be examined – background noise at the site was low frequency. The system was able to pinpoint accurately the pipe fracture. Further tests confirmed other leakages, including one of 200 L min^{-1}, and considerable savings have been made. Even good results on plastic pipes have been achieved.

13.9 Pipe line corrosion (Real Time Corrosion Management, undated)

Condition and detection

All offshore workers are aware of the seriousness of corrosion, whether in oil, gas or water flow lines. It is for this reason that it is normal practice to apply corrosion inhibitors in a continuous manner to prevent, or at least to reduce, the corrosion. This can be either too little – and pointless – or too much – and excessively expensive. CML were therefore appointed to monitor the corrosion rate against the level of inhibitor, to ensure the optimum amount was injected.

Result

It was assessed that the uninhibited corrosion rate was 2.0 mm per year. Continuous injection was then applied at 30 ppm. In 1.5 h, the corrosion rate had dropped to 30% and still further to 20% within 13 h, equivalent to 0.4 mm per year.

13.10 Pump efficiency (Yates, 1994)

Condition and detection

Electrically driven pumps cost approximately £400 per kilowatt per year to run. Energy can be saved by improving the efficiency of the pump according to the equation

$$\text{percentage saving} = (1 - \eta_1/\eta_2) \times 100\%$$

where η_1 is the initial efficiency and η_2 the improved efficiency. By using an efficiency meter, such as the Yatesmeter, the initial and running efficiency has been examined in many situations (Fig. 13.4).

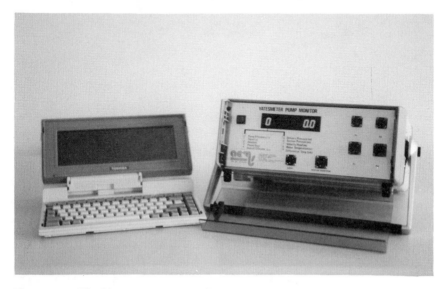

Figure 13.4 The Yatesmeter pump efficiency monitor.

Result

One particular example cited was that of a 500 kW machine which was checked by means of the Yatesmeter as having an efficiency of 65%. This was considered far too low, and, by refurbishing the pump, an improvement to 82% was achieved. This meant a percentage saving of

$$(1 - 65 / 82) \times 100\% = 20.73\%$$

and the annual saving on the pump of

$$£400 \, \text{kW}^{-1} \times 500 \, \text{kW} \times 0.2073 = £41,460.$$

The refurbishment cost had been £15,000, which had then been paid off in the efficiency saving in little more than four months.

PART FOUR

Appendices

Tables and charts

There are numerous examples in table form in the main part of the book. In addition there are some tables which can be used for checking such as Table 3.3 (checking for possibilities of failure), Table 4.2 (checking for possibilities of monitoring) and Table 5.1 (checking for overall costs with monitoring).

Table A.1 Monitoring features to be considered

Number	Monitoring feature	Comments	?
1	Able to monitor all faults	Not very likely	
2	Easy to fit	Unless it is easy to fit, it will possibly never leave its box	
3	Easy to use	Imagine using it many times; is it still easy?	
4	More reliable than the system being tested	The supplier will not admit to unreliability; check with other users	
5	Able to be checked and calibrated	How often need this to be done? What is the cost of calibration or checking?	
6	Able to give an immediate answer	How soon is an answer really required?	
7	Able to be compensated for variables	Variables such as people, atmospherics, machine conditions	
8	Adaptable to different systems	Can it be used in other applications? (a possible cost saving)	
9	Able to store evidence for later checking	Always beneficial in any dispute, or if further analysis required	
10	Inexpensive	In comparison with the overall cost of the machine operation (include the running cost as well as purchase price)	
11	Small and light	What size is acceptable? What size is preferable?	
12	Non-iatrogenic	Will not cause a different sort of failure because it has been fitted	
13	Safe	Electrically, chemically, mechanically and hydraulically	
14	Economically viable	10 above compared with cost of failure	
15	Easy to obtain and maintain	Applies to spares as well as the original purchase	

Used in Table 2.5. The ? column allows an importance rating to be inserted.

In this appendix there are also the following charts and tables:

Table A.2 Check list for possible monitoring methods (Table 11.3)

Type of monitor	Components considered for monitoring	Monitor preference	✓
Environment:			
Temperature			
Relative humidity			
Pressure			
Light			
Vibration			
Air velocity			
Pollution			
Radiation			
Gas			
Power			
Other (input)			
Human:			
Colour			
Smell			
Vibration			
Ultrasonics			
Wear debris			
Oil			
Thermography			
Leakage			
Corrosion			
Steady state:			
Pressure			
Flow			
Level			
Speed			
Temperature			
Other (within)			
Performance			
Product variants			
Other (output)			

Possibility rating: 1, not very likely; 2, just a possibility; 3, a possibility; 4, 50:50; 5, more than just likely; 6, likely; 7, very likely; 8, highly likely indeed.

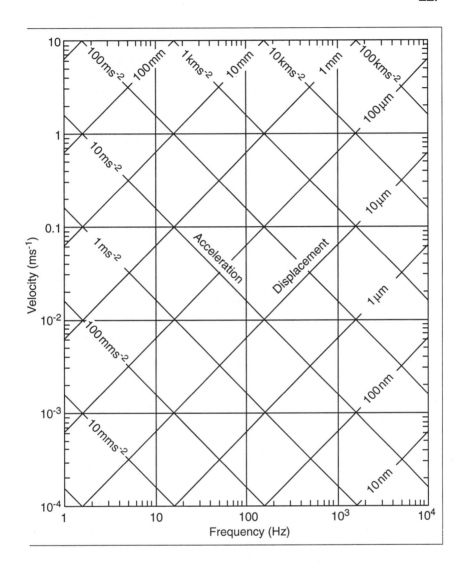

Figure A.1 Frequency, displacement velocity and acceleration.

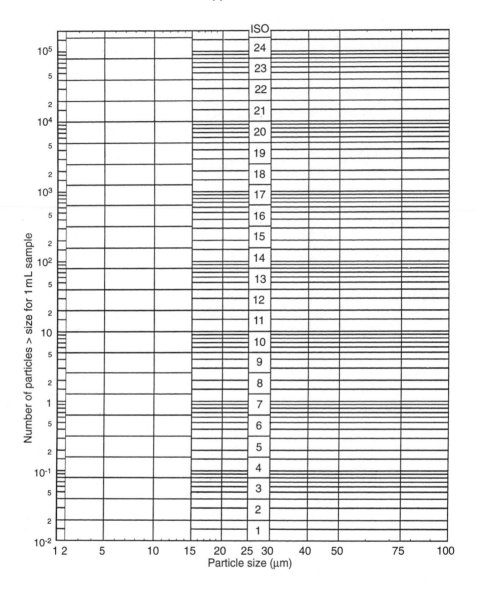

Figure A.2 ISO 4406 chart for assessing contamination level. The standard is quoted as a three-figure number, e.g. 20/15/12, denoting the ISO coding bands for the particle counts at 2μm, 5μm and 15μm per 1 mL sample size. A dash implies that no count was made at that size, e.g. –/15/12. An asterisk (*) implies that there were too many particles to count.

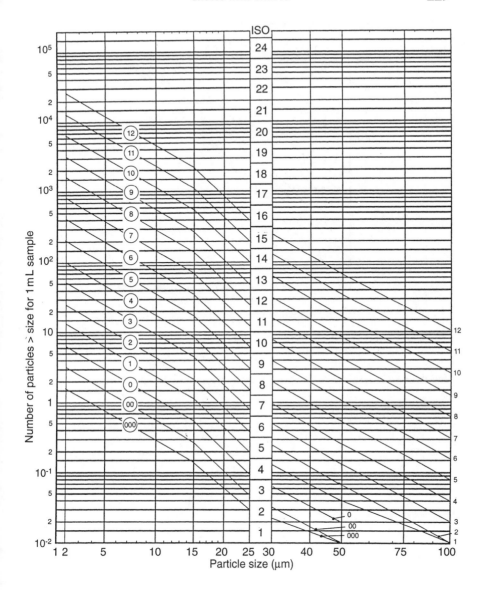

Figure A.3 NAS 1638 chart for assessing contamination level. The particle distribution is drawn on the chart and the lowest NAS code taken which is above all the points in the distribution. Strictly speaking, the distribution should cover the range from 5µm to 100µm (and possibly down to 2µm) but this is not always done.

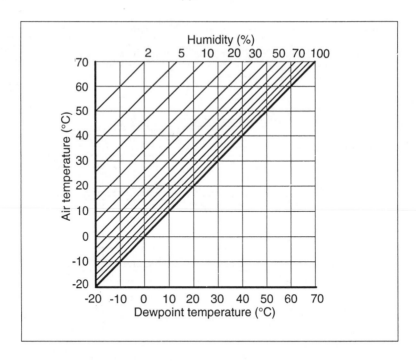

Figure A.4 Assessment of dewpoint taking into account temperature and humidity.

1 H																	2 He
3 Li	4 Be											5 B	6 C	7 N	8 O	9 F	10 Ne
11 Na	12 Mg											13 Al	14 Si	15 P	16 S	17 Cl	18 Ar
19 K	20 Ca	21 Sc	22 Ti	23 V	24 Cr	25 Mn	26 Fe	27 Co	28 Ni	29 Cu	30 Zn	31 Ga	32 Ge	33 As	34 Se	35 Br	36 Kr
37 Rb	38 Sr	39 Y	40 Zr	41 Nb	42 Mo	43 Tc	44 Ru	45 Rh	46 Pd	47 Ag	48 Cd	49 In	50 Sn	51 Sb	52 Te	53 I	54 Xe
55 Cs	56 Ba	57* La	72 Hf	73 Ta	74 W	75 Re	76 Os	77 Ir	78 Pt	79 Au	80 Hg	81 Tl	82 Pb	83 Bi	84 Po	85 At	86 Rn
87 Fr	88 Ra	89** Ac	104 (Ku)	105 (Ns)													

* Lanthanide

58 Ce	59 Pr	60 Nd	61 Pm	62 Sm	63 Eu	64 Gd	65 Tb	66 Dy	67 Ho	68 Er	69 Tm	70 Yb	71 Lu

** Actinide

90 Th	91 Pa	92 U	93 Np	94 Pu	95 Am	96 Cm	97 Bk	98 Cf	99 Es	100 Fm	101 Md	102 Md	103 Lr

Figure A.5 The Periodic Table (elements up to number 105).

Definitions

There is a wide range of acronyms, synonyms and buzz words frequently used in condition monitoring. While some may appear obvious, the subtle undertones in the meaning of some may not be fully known. These brief amplifications of initials, with the occasional description, should put the reader right. For other definitions please see the Index.

ADC	Analogue-to-digital (AD) converter
AE	Acoustic emission
AFNOR	Association Français de Normalisation (French standards association)
AI	Artificial intelligence
ANN	Artificial neural network (neural network – the standard name for the computer-based neural network)
ANS	Adaptive network system
ANSI	American National Standards Institute
ASCII	American Standard Code for Information Interchange
ASME	American Society of Mechanical Engineers
ASTM	American Society for Testing and Materials
BFPA	British Fluid Power Association
BInstNDT	British Institute of Non-destructive Testing
BSI	British Standards Institution
CAD	Computer-aided design
CAM	Computer-aided manufacture
CAMM	Computer-aided maintenance management
CAT	Computer-aided testing
CBM	Computer-based maintenance
CEN	Comité Européen de Normalisation (European Standards Committee)

CME	Component–monitor–economics (the analysis process for monitoring)
CMMS	Computer maintenance management system
CMSA	Condition Monitoring and Sound Assessment Club (set up in 1993 to examine the use of neural networks for evaluating machine condition and the subjective noise generated)
CPU	Central processing unit (maths processing within a computer)
CVD	Chemical vapour deposition (used in the manufacture of pressure sensors)
DAC	Digital-to-analogue (DA) converter
DAPS	Data acquisition and processing system
DCS	Distributed control systems
DIN	Deutsche Institut für Normung (German standards authority)
DMA	Direct memory access
DSP	Digital signal processing
EEPROM	Electrically EPROM (erased by electrical signal)
EIS	Electrochemical impedance spectroscopy (corrosion)
EISA	Extended Industry Standard Architecture (see ISA[2])
EMC	Electromagnetic compatibility
EMI	Electromagnetic interference
EN	Electrochemical noise (corrosion)
EPROM	Erasable programmable read-only memory (erased by UV light)
ESP[1]	Envelope signal processing (vibration)
ESP[2]	Extra sensory prediction (predictive maintenance)
FFT	Fast Fourier transform
FIM	Flexible interface module
FM	Frequency modulated
FMA	Failure mode analysis (similar to CME but with more maths)
FSM	Field signature method (corrosion)
FTIR	Fourier transform infra-red (spectrometric oil analysis)
GPIB	General purpose interface bus (data acquisition)
HA	Harmonic analysis (corrosion)
HMI	Human–machine interface
IEE	Institute of Electrical Engineers

IEEE488 etc.	Standard computer interfaces (see Index)
IR	Infra-red
ISA[1]	International Society for Measurement and Control
ISA[2]	Industry standard architecture (computer 'bus' or line down which data flows)
ISO	International Standards Organization
K	1024 (2^{10}) when related to number of computer bytes (approximately A5 text)
KBS	Knowledge-based system
LAN	Local area network (PCs joined together locally and sharing information)
LCD	Liquid crystal display
LED	Light emitting diode
LPRM	Linear polarization resistance measurements
LVDT	Linear variable differential transformer (steady state monitoring)
MIS	Management information system
MLP	Multilayer perceptron (neural network – a particular neural function which uses several layers of neurons, where each neuron takes several inputs and sums them in the course of producing an output)
MMI	Man–machine interface
MPI	Magnetic particle inspection (corrosion – crack inspection)
MTBF	Mean time between failures
MTTF	Mean time to failure
MTTR	Mean time to repair
NAMAS	National Measurement Accreditation Service
NAS	National Aerospace Standard (USA)
NDIR	Non-dispersive infra-red (measurement of gas content)
NDT	Non-destructive testing
NFPA	National Fluid Power Association (USA)
NO_x	Nitrogen oxides
NRB	Nitrite-reducing bacteria (corrosion)
PC	Personal computer
PCI	Peripheral component interconnect (a standard for the PC interface with a theoretical bandwidth at 132 MBytes s^{-1} (some ten times greater than the previous ISA bus)

PCMCIA	Personal Computer Memory Card International Association – now renamed 'PC Card' (a credit card sized device which slots into a PC to give an efficient means of extending communications (telephone, fax, video, voice and paging capabilities))
PID	Proportional integral derivative
PLC	Programmable logic controller (factory operations are monitored and rapidly controlled through the rapid monitoring of numerous parameters such as pressure, temperature etc.)
PROM	Programmable read-only memory
RAM	Random access memory (store in PC where worked-on data resides)
RBF	Radial basis function (neural network – a particular function used within a hidden layer neuron)
RBM	Reliability-based maintenance
RCM	Reliability-centred maintenance
RFI	Radio frequency interference
RGB	Red, green, blue (primary colours)
RMS	Root mean square
ROM	Read-only memory (data for retrieval only)
RPD	Rotary particle depositor (wear analysis instrument based on ferrography)
RS232, etc.	Standard computer interfaces (see Index)
SAE	Society of Automotive Engineers (USA)
SCADA	Supervisory control and data acquisition (used for factory area supervision and control including several PLCs)
SEE[1]	Society of Environment Engineers
SEE[2]	Spectral emitted energy (ultrasonics)
SEM	Scanning electron microscopy
SO_x	Sulphur oxides
SPC	Statistical process control
SPM	Shock pulse meter (ultrasonics)
SRB	Sulphur-reducing bacteria (corrosion)
TMH	Thermo-magnetic hysteresis
TOC	Total organic carbon
TOD	Total organic demand
TPM	Total productive maintenance
TQM	Total quality maintenance

VDU	Visual display unit (like a PC with a television screen)
VESA	Video Electronics Standards Association (computer process bus in 1992)
VI	Virtual instrument (the PC behaves like a conventional instrument)
VOC	Volatile organic compounds
VXI	VME extensions for instrumentation
ZRA	Zero resistance ammetry (corrosion)

Condition monitoring companies

The company names and addresses are of those who have contributed either directly or indirectly to the contents of this book. They are not necessarily approved or even considered as better than others which have not been included, but they do allow the reader to further his or her knowledge of what is available for monitoring. The product listing is not complete, but it gives an introduction to many of the relevant items and services which relate to condition monitoring. Please contact the companies for further details.

Key:
A – Colour
B – Corrosion
C – Communication and data acquisition
D – Environment (air flow, RH etc.)
E – Flow
F – Leakage
G – Level
H – Maintenance and management
J – Motion (speed, rotation, displacement)
K – Neural networks
L – Oil (or liquid) analysis
M – Performance and efficiency
N – Power, torque, strain, load
P – Pressure (in system)
Q – Sampling (of liquids)
R – Smell and gas (not air)
S – Temperature and thermography
T – Training (in one or more parts of condition monitoring)
U – Ultrasonics, stress, AE
V – Vibration, noise
W – Wear debris analysis
X – Water content

A

ABB Kent-Taylor Ltd, Howard Road, Eaton Socon,   ACEFGHPRS
St. Neots, Huntingdon, Cambs, PE19 3EU;
Tel: 01480 475321, Fax: 01480 217948

Able Instruments & Controls Ltd, Cutbush Park,   DEGLPRSQ
Danehill, Lower Early, Reading, Berks, RG6 4UT;
Tel: 01734 311188, Fax: 01734 312161

Acal Auriema Ltd, 442 Bath Road, Slough,   DEGKMNQSX
Bucks, SL1 6BB;
Tel: 01628 604353, Fax: 01628 603730

Adept Scientific plc, 6 Business Centre West,   C
Avenue One, Letchworth, Herts, SG6 2HB;
Tel: 01462 480055, Fax: 01462 480213

Advanced Energy Monitoring Systems Ltd,   MV
The Energy Centre, Finnimore Industrial
Estate, Ottery St. Mary, Devon, EX11 1NR;
Tel: 01404 812294, Fax: 01404 812603

Advanced Expert Systems Ltd, Bio House,   H
Derwent Street, Derby, DE1 2EF;
Tel: 01332 383521, Fax: 01332 383532

AEA Sonomatic Ltd, 20 Rivington Court,   BU
Hardwick Grange, Woolston, Warrington, WA1 4RT;
Tel: 01925 810511, Fax: 01925 828699

AEA Technology, Risley, Warrington,   BDHKLQSUX
Cheshire, WA3 6AT;
Tel: 01925 252971, Fax: 01925 252579

Agema Infrared Systems Ltd, Arden House,   S
West Street, Leighton Buzzard, Beds, LU7 7DD;
Tel: 01525 375660, Fax: 01525 379271

Agma plc, Gemini Works, Haltwhistle,   BF
Northumberland, NE49 9JW;
Tel: 01434 320598, Fax: 01434 321650

AI Cambridge Ltd, London Road, Pampisford,   F
Cambridge, CB2 4EF;
Tel: 01223 495550, Fax: 01223 835050

Airflow Developments Ltd, Cressex Business Park,   CDS
Lancaster Road, High Wycombe, Bucks, HP12 3QP;
Tel: 01494 525252, Fax: 01494 461073

A.J. Thermosensors Ltd, Martlets Way, Goring by Sea, DS
Worthing, West Sussex, BN12 4HF;
Tel: 01903 502471, Fax: 01903 506231

Allen Bradley Industrial Automation Products,   GJPS
Pitfield, Kiln Farm, Milton Keynes, MK11 3DR;
Tel: 01908 838800, Fax: 01908 261917

Allison Engineering Ltd, Allison House,   EGMPR
    Cranes Farm Road, Basildon, Essex, SS14 3JA;
    Tel: 01268 526161, Fax: 01268 533144
Alpine Components (UE Systems), PO Box 3,   CFU
    Battle, East Sussex, TN33 0XW;
    Tel: 01424 775451, Fax: 01424 775010
Alrad Instruments Ltd, Alder House, Turnpike Road   RS
    Industrial Estate, Newbury, Berks, RG13 2NS;
    Tel: 01635 30345, Fax: 01635 32630
Amot Controls, Western Way, Bury St. Edmunds,   CGJPS
    Suffolk, IP33 3SZ;
    Tel: 01284 762222, Fax: 01284 760256
Analytical Development Company Ltd, Pindar Road,   DR
    Hoddesdon, Herts, EN11 0AQ;
    Tel: 01992 469638, Fax: 01992 444567
Applied Imaging International Ltd, Dukesway,   W
    Team Valley, Gateshead, NE11 0PZ;
    Tel: 0191 482 2111, Fax: 0191 482 5249
AromaScan plc, Electra House, Electra Way,   R
    Crewe, CW1 1WZ;
    Tel: 01270 216444, Fax: 01270 216030
Ashdown Process Control Ltd, Unit 3, North Close,   G
    Shorncliffe Industrial Estate, Folkestone, Kent,
    CT20 3UH;
    Tel: 01303 240691, Fax: 01303 240703
ASM UK Ltd, 18A Firtree Lane, Groby,   CJPS
    Leics, LE6 0FH;
    Tel: 0116 287 4447, Fax: 0116 287 2265
ATL Consulting Services Ltd, 36–38 The Avenue,   H
    Southampton, SO17 1XN;
    Tel: 01703 325000, Fax: 01703 335251
Automatic Systems Laboratories Ltd,   MS
    28 Blundells Road, Bradville, Milton Keynes,
    MK13 7HF;
    Tel: 01908 320666, Fax: 01908 322564

B

Baird Europe BV, Postbus 81, Produktieweg 30,   LWX
    NL-2328 AB, Zoeterwoude, Netherlands;
    Tel: 0071 413151, Fax: 0071 414899
Belstock Controls, 10 Moss Hall Crescent,   DLW
    Finchley, London, N12 8NY;
    Tel: 0181 446 8210, Fax: 0181 446 6991

Bently & Nevada (UK) Ltd, 2 Kelvin Close,   V
    Science Park North, Birchwood, Warrington,
    Cheshire, WA3 7PB;
    Tel: 01925 818504, Fax: 01925 817819
Biodata Ltd, 10 Stocks Street, Manchester M8 8QG;   B
    Tel: 0161 834 6688, Fax: 0161 833 2190
BIRAL (TSI), Bristol Industrial & Research   DEJPR
    Associates Ltd, PO Box 2, Portishead, Bristol,
    BS20 9JB;
    Tel: 01275 847787, Fax: 01275 847303
Bourdon Sedeme Ltd, Horton Close, West Drayton,   NPS
    Middx, UB7 8JA;
    Tel: 0189 432832, Fax: 0189 432834
BP Oil UK Ltd, BP House, Breakspear Way,   LW
    Hemel Hempstead, Herts, HP2 4UL;
    Tel: 01442 232323, Fax: 01442 225509
British Fluid Power Association, Cheriton House,   HT
    Cromwell Business Park, Banbury Road,
    Chipping Norton, Oxon, OX7 5SR;
    Tel: 01608 644114, Fax: 01608 643738
Brookfield Viscometers Ltd, 1 Whitehall Estate,   L
    Flex Meadow, Pinnacles West, Harlow,
    Essex, CM19 5YF;
    Tel: 01279 451774, Fax: 01279 451775
Brookhaven Instruments Ltd, Chapel House,   KW
    Stock Wood, Worcs, B96 6ST;
    Tel: 01386 792727, Fax: 01386 792727
Brownell Ltd, Helena Works, Mordaunt Road,   D
    London, NW10 8PX;
    Tel: 0181 965 9281, Fax: 0181 965 3239
Brüel & Kjær (UK) Ltd, Harrow Weald Lodge,   CDRV
    92 Uxbridge Road, Harrow, Middx, HA3 6BZ;
    Tel: 0181 954 2366, Fax: 0181 954 9504
Budenburg Gauge Co. Ltd, PO Box 5, Woodfield Road,   P
    Broadheath, Altrincham, WA14 4ER;
    Tel: 0161 928 5441, Fax: 0161 928 7075

C
Cadar, Fernie Road, Market Harborough,   L
    Leics, LE16 7PH;
    Tel: 01858 410101, Fax: 01858 433934
Calex Instrumentation Ltd, PO Box 2,   S
    Leighton Buzzard, Beds, LU7 8WZ;
    Tel: 01525 373178, Fax: 01525 851319

Cambridge Control Solutions, Newton House,   K
Cambridge Business Park, Cowley Road,
Cambridge, CB4 4WZ;
Tel: 01223 423200, Fax: 01223 423255

Cambridge Monitoring Systems Ltd, St. Johns   C
Innovation Centre, Cowley Road, Cambridge,
CB4 4WS;
Tel: 01223 276791, Fax: 01223 277436

Camlab Ltd, Nuffield Road, Cambridge, CB4 1TH;   LQWX
Tel: 01223 424222, Fax: 01223 420856

Canongate Technology Ltd, 36 Inglis Green Road,   C
Edinburgh, EH14 2ER;
Tel: 0131 455 7211, Fax: 0131 455 7928

Casella London Ltd, Regent House, Wolseley Road,   DR
Kempston, Beds, MK42 7JY;
Tel: 01234 841441, Fax: 01234 841490

Castrol (UK) Ltd, Industrial Division, Pipers Way,   LWX
Swindon, Wilts, SN3 1RE;
Tel: 01793 512712, Fax: 01793 486083

Century Oils Ltd, PO Box 2, New Century Street,   LWX
Hanley, Stoke-on-Trent, ST1 5HU;
Tel: 01782 202521, Fax: 01782 202073

Chell Instruments Ltd, Tudor House, Grammar   ADEFMPR
School Road, North Walsham, Norfolk, NR28 9JH;
Tel: 01692 402488, Fax: 01692 406177

Chryseptre (UK) Ltd, Studio 2, Highfield Farm Court,   C
Huncote Road, Stoney Stanton, Leics, LE9 6DJ;
Tel: 01455 271707, Fax: 01455 271172

C-Matic Systems Ltd, The Forge, Park Road,   N
Crowborough, East Sussex, TN6 2QX;
Tel: 01892 665688, Fax: 01892 667515

CML – see Real Time Corrosion Management

Col-Ven S.A., Ruta 11 Km 814, 3574   FP
Guadalupe Norte, Santafe, Argentina;
Tel: 0054 482 98000, Fax: 0054 482 98030

Cole-Parmer Instrument Co Ltd, PO Box 22,   BDEFGJPSX
Bishop's Stortford, Herts, CM23 3DX;
Tel: 01279 757711, Fax: 01279 755785

Compact Instruments Ltd, Binary Works, Park Road,   J
Barnet, Herts, EN5 5SA;
Tel: 01440 6663, Fax: 01440 9956

Control & Readout Ltd, Stafford Park 6, Telford,   CS
Shropshire, TF3 3BQ;
Tel: 01952 292527 Fax: 01952 292654

Copper Development Association, Orchard House,   MN
     Mutton Lane, Potters Bar, Herts, EN6 3AP;
     Tel: 01707 650711, Fax: 01707 642769

CorrOcean Ltd, 430 Clifton Road, Aberdeen, AB2 2EJ;   B
     Tel: 01224 662180, Fax: 01224 662070

Coulter Electronics Ltd, Northwell Drive,   LW
     Luton, Beds, BN15 8AJ;
     Tel: 01582 567000, Fax: 01582 490390

Crane Perflow Ltd, Unit 3, Chapmans Park Industrial   E
     Estate, High Road, Willesden, London, NW10 2DY;
     Tel: 0181 451 4577, Fax: 0181 451 6788

Crowcon Detection Instruments Ltd,   R
     2 Blacklands Way, Abingdon Business Park,
     Abingdon, Oxon, OX14 1DY;
     Tel: 01235 553057, Fax: 01235 553062

CSI (UK) Ltd, Unit 2, Deeside Enterprise Centre,   HLNSV
     Rowleys Drive, Deeside, Clwyd, CH5 1PP;
     Tel: 01244 822115, Fax: 01244 823100

D

Danfoss Flowmetering Ltd, Magflow House,   EG
     Ebley Road, Stonehouse, Glos, GL10 2LU;
     Tel: 01453 828891, Fax: 01453 824013

Data Acquisition Ltd, Electron House,   C
     Higher Hillgate, Stockport, Cheshire, SK1 3QD;
     Tel: 0161 477 3888, Fax: 0161 480 5142

Data Cell Ltd, Hattori House, Vanwall Business   CLS
     Park, Maidenhead, Berks, SL6 4UB;
     Tel: 01628 415415, Fax: 01628 415400

Data Electronics, Business Centre West, Avenue One,   C
     Letchworth, Herts, SG6 2HB;
     Tel: 01462 481291, Fax: 01462 481375

Data Translation Ltd, The Mulberry Business Park,   C
     Wokingham, Berks, RG11 2QJ;
     Tel: 01734 793838, Fax: 01734 776670

De la Pena Ltd, Racecourse Road, Pershore,   L
     Worcs, WR10 2DD;
     Tel: 01386 552311, Fax: 01386 556401

Delta Technical Services Ltd, Asser House,   DFR
     Airport Service Road, Portsmouth, Hants, PO3 5RA;
     Tel: 01705 697321, Fax: 01705 673668

Delta-Technical Devices Ltd, 128 Low Road,   D
     Burwell, Cambs, CB5 0EJ;
     Tel: 01480 495047, Fax: 01480 493310

Detectaids (UK) Ltd, Singleton Court,   FU
   Wonastow Road, Monmouth, Gwent, NP5 3AH;
   Tel: 01600 716911, Fax: 01600 772976

Detectawl (Gastec) Ltd, 2 Cochran Close, Crownhill,   R
   Milton Keynes, Bucks, MK8 0AJ;
   Tel: 01908 568076, Fax: 01908 260593

Diagnostic Instruments Ltd,   CKV
   2 Michaelson Square, Kirton Campus,
   Livingston, EH54 7DP;
   Tel: 01506 470011, Fax: 01506 470012

Diagnostic & Measuring Systems Ltd,   HNV
   9/10 Bridgend Road, Dingwall, Ross-shire IV15 9SL;
   Tel: 01349 861048, Fax: 01349 861394

Dickinson Control Systems Ltd, Unit E,   C
   Moorside Road, Winchester, Hants, SO23 7RX;
   Tel: 01962 840333, Fax: 01962 842011

Digitron Instrumentation Ltd, Technology House,   CDJMPRS
   Mead Lane, Hertford, SG13 7AW;
   Tel: 01992 587441, Fax: 01992 500028

Draeger Ltd, Ullswater Close, Kitty Brewster   R
   Industrial Estate, Blyth, Northumberland,
   NE24 4RG;
   Tel: 01670 352891, Fax: 01670 356266

Dresser UK Instruments Division, Rufford Court,   EG
   Hardwick Grange, Warrington, Cheshire,
   WA12 4RF;
   Tel: 01925 814545, Fax: 01925 816378

Druck Ltd, Fir Tree Lane, Groby, Leicester, LE6 0FH;   PSV
   Tel: 0116 231 4314, Fax: 0116 231 4192

Dwyer Instruments Ltd, Unit 16, The Wye Estate,   EGPS
   London Road, High Wycombe, Bucks, HP11 1LH;
   Tel: 01494 461707, Fax: 01494 465102

Dynamic Logic Ltd, The Western Centre,   CGH
   Western Road, Bracknell, Berks, RG12 1RW;
   Tel: 01344 51915, Fax: 01344 52253

E

Elcomponent Ltd, Unit 5, Southmill Trading Centre,   MN
   Bishop's Stortford, Herts, CM23 3DP;
   Tel: 01279 503173, Fax: 01279 654441

Electronic Temperature Instruments Ltd,   CS
   Southdown View Road, Broadwater Trading
   Estate, Worthing, West Sussex, BN14 8NL;
   Tel: 01903 202151, Fax: 01903 202445

Endevco UK Ltd, Melbourn, Royston, Herts,   CJPV
    SG8 6AQ;
    Tel: 01763 261311, Fax: 01763 261120

Endress + Hauser Ltd, Ledson Road,   CDEGNP
    Manchester, M23 9PH;
    Tel: 0161 998 0321, Fax: 0161 998 1841

Engica Technology Systems International, Newcastle   H
    Technopole, Central Business and Technology Park,
    Kings Manor, Newcastle upon Tyne, NE1 6PA;
    Tel: 0191 201 7777, Fax: 0191 201 7778

Entek Scientific Corp., The Maltings, Charlton Road,   HV
    Shepton Mallet, Somerset, BA4 5QE;
    Tel: 01749 344878, Fax: 01749 346285

Entran Ltd, 19 Garston Park Parade, Garston,   NPSV
    Watford, Herts, WD2 6LQ;
    Tel: 01923 893999, Fax: 01923 893434

Environmental Monitoring Systems Ltd,   DLW
    G11 Mayford Business Centre, Smarts Heath Road,
    Woking, Surrey, GU22 0PP;
    Tel: 01483 722463, Fax: 01483 740462

Epic Products Ltd, Pacific Way, Salford,   Q
    Manchester, M5 2DL;
    Tel: 0161 872 1487, Fax: 0161 848 7324

Erwin Sick Optic-Electronic Ltd, Waldkirch House,   ADEJQRU
    39 Hedley Road, St Albans, Herts, AL1 5BN;
    Tel: 01727 831121, Fax: 01727 856767

Eternit Pipes Division, Meridian House, Park Road,   F
    Swanley, Kent, BR8 8AH;
    Tel: 01322 614414, Fax: 01322 666734

Eurogauge Company Ltd, Imberhorne Lane,   EG
    East Grinstead, West Sussex, RH19 1RF;
    Tel: 01342 323641, Fax: 01342 315513

F

Fenner Electronic Controls, J.H. Fenner Ltd,   N
    Mount View Works, Whitcliffe Road,
    Cleckheaton, West Yorks, BD19 3AG;
    Tel: 01274 851234, Fax: 01274 851413

Fisher-Rosemount Ltd, Heath Place, Bognor Regis,   EKHPS
    West Sussex, PO22 9SH;
    Tel: 01243 863121, Fax: 01243 867554

Flowdata Systems Ltd, PO Box 71, Newbury,   EG
    Berks, RG14 7JZ;
    Tel: 01635 521066, Fax: 01635 33741

Flowline Manufacturing Ltd, 11a Shenley Road,   EG
     Borehamwood, Herts, WD6 1AD;
     Tel: 0181 207 6565, Fax: 0181 207 3082

Fluke (UK) Ltd, Colonial Way, Watford, Herts, WD2 4TT;   CNPSV
     Tel: 01923 240511, Fax: 01923 225067

Fluid Power Centre, University of Bath,   T
     Claverton Down, Bath, Avon, BA2 7AY;
     Tel: 01225 826373, Fax: 01225 826928

Flupac Ltd, Woodstock Road, Charlbury, Oxon,   EGLPW
     OX7 3EF;
     Tel: 01608 811211, Fax: 01608 811259

FMA Ltd, FMA House, Hogwood Lane,   ELPS
     Finchampstead, Wokingham, Berks, RG40 4QW;
     Tel: 01734 730100, Fax: 01734 328094

Fulmer – see IPH Fulmer

G

GE Fanuc Automation (UK) Ltd, Unit 1, Mill Square,   H
     Featherstone Road, Wolverton Mill South,
     Milton Keynes, MK12 5BZ;
     Tel: 01908 226488, Fax: 01908 226551

Gelman Sciences Ltd, 11 Harrowden Road,   LW
     Brackmills, Northampton, NN4 0EZ;
     Tel: 01604 765141, Fax: 01604 761383

Gensym Ltd, Kings Chase, 107 King Street,   K
     Maidenhead, Berks, SL6 1DP;
     Tel: 01628 788661, Fax: 01628 37999

Geotechnical Instruments (UK) Ltd,   FGPQR
     Sovereign House, Queensway, Leamington Spa,
     Warwicks, CV31 3JR;
     Tel: 01926 338110, Fax: 01926 338110

GoRaTec (UK) Ltd, 47 Cavendish Road,   S
     Ellesmere Park, Eccles, Manchester, M30 9EE;
     Tel: 0161 788 9929, Fax: 0161 788 9930

Grant Instruments (Cambridge) Ltd, Barrington,   CS
     Cambridge, CB2 5QX;
     Tel: 01763 262600, Fax: 01763 262410

Graseby, Park Avenue, Bushey, Herts, WD2 2BW;   DQR
     Tel: 01923 228566, Fax: 01923 240285

H

Hawco Ltd, Cathedral Hill Industrial Estate,   CDGJPS
     Guildford, Surrey, GU2 5YB;
     Tel: 01483 60606, Fax: 01483 575973

Hewlett Packard Ltd, Cain Road, Bracknell,   CV
   Berks, RG12 1HN;
   Tel: 01344 360000, Fax: 01344 363344
HMD Seal/Less Pumps Ltd, Hampden Park Ind.   HNS
   Estate, Eastbourne, East Sussex, BN22 9AN;
   Tel: 01323 501241, Fax: 01323 503369
Hohner Automation Ltd, Whitegate Road, Wrexham,   R
   Clwyd, LL13 8RB;
   Tel: 01978 264888, Fax: 01978 364586
Holroyd Instruments, Unit 308,Via Gellia Mills,   FU
   Bonsall, Matlock, Derby, DE4 2AJ;
   Tel: 01629 822060 (also fax)
Honeywell Control Systems Ltd, Honeywell House,   A
   Arlington Business Park, Bracknell, Berks,
   RG12 1EB;
   Tel: 01344 826000, Fax: 01344 826240
Howden Wade Ltd, Thermal Control Products,   LW
   Crowhurst Road, Brighton, East Sussex, BN1 8AJ;
   Tel: 01273 506311, Fax: 01273 557123
Humitec Ltd, Longley House, East Park,   CD
   Crawley, West Sussex,
   RH10 6AD;
   Tel: 01293 564321, Fax: 01293 564730
Hydrotechnik UK Ltd, Unit 4, 55A Yeldham Road,   ELQW
   Hammersmith, London, W6;
   Tel: 0181 741 9934, Fax: 0181 741 9935

I

INA Bearing Co. Ltd, Forge Lane, Minworth,   UV
   Sutton Coldfield, West Midlands, B76 8AP;
   Tel: 0121 351 3833, Fax: 0121 351 7686
Industrial Flow Control Ltd, 30 Hornsby Square,   E
   Southfields Industrial Estate, Laindon,
   Essex, SS15 6SD;
   Tel: 01268 540429, Fax: 01268 541270
Insight Vision Systems Ltd, Unit 6, Merebrook,   FSU
   Hanley Road, Malvern, Worcs, WR13 6NP;
   Tel: 01684 310001, Fax: 01684 310510
Intec Controls Ltd, PO Box 33, Chichester,   CH
   West Sussex, PO20 7DW;
   Tel: 01243 532763, Fax: 01243 530672
Intercontrol Ltd, PO Box 29, Gainsborough,   EGPS
   Lincs, DN21 5LR;
   Tel: 01427 610794, Fax: 01427 615976

Ion Science Ltd, The Way, Fowlmere, Cambridge,   F
    SG8 7QP;
    Tel: 01763 208503, Fax: 01763 208814
Ionics UK Ltd, Unit 3, Mercury Way, Mercury   LR
    Park Estate, Urmston, Manchester, M41 7LY;
    Tel: 0161 866 9337, Fax: 0161 866 9630
IPH Fulmer Ltd, Barn Close, Old Lane, Farthinghoe,   LW
    Northants, NN13 5NZ;
    Tel: 01295 712551, Fax: 01295 712551
Ircon Ltd, Unit 6, Park Road, Swanley, Kent,   JMS
    BR8 8AH;
    Tel: 01322 613224, Fax: 01322 613328
IRD Mechanalysis (UK) Ltd, TEC Product Division,   CTV
    Bumpers Lane, Sealand Industrial Estate,
    Chester, CH1 4LT;
    Tel: 01244 374914, Fax: 01244 379870
ITT Barton UK, 3 Steyning Way, Southern Cross   BDEGPS
    Trading Estate, Bognor Regis, West Sussex,
    PO22 9TT;
    Tel: 01243 826741, Fax: 01243 860263

K
KDG Mobrey Ltd, 190–196 Bath Road, Slough,   EGHPS
    Berks, SL1 4DN;
    Tel: 01753 534646, Fax: 01753 823589
Keithley Instruments Ltd, The Minster,   C
    58 Portman Road, Reading, Berks, RG3 1EA;
    Tel: 01734 575666, Fax: 01734 596469
Keller (UK) Ltd, Alpha House, 79 High Street,   P
    Crowthorne, Berks, RG11 7AD;
    Tel: 01344 780199, Fax: 01344 778916
Kingston Condition Monitoring Services Ltd,   SV
    1A Lorraine Street, Stoneferry Road, Hull, HU8 8EG;
    Tel: 01482 227953, Fax: 01482 587247
Kistler Instruments Ltd, Whiteoaks, The Grove,   NPV
    Hartley Witney, Hants, RG27 8RN;
    Tel: 01252 843555, Fax: 01252 844439
Krohne Measurement & Control Ltd, Rutherford Drive,   EGPS
    Park Farm Industrial Estate, Wellingborough,
    Northants, NN8 6AE;
    Tel: 01933 408500, Fax: 01933 408501
KT Hydraulic Systems Ltd, Hope Hall Mill,   E
    Union Street South, Halifax, West Yorks, HX1 2LA;
    Tel: 01422 358885, Fax: 01422 359512

Kulite Sensors Ltd, Stroudley Road, Kingsland   NPSV
    Business Park, Basingstoke, Hants, RG24 8UG;
    Tel: 01256 461646, Fax: 01256 479510

L

Labfacility Ltd, 99 Waldegrave Road, Teddington,   CMNPS
    Middx, TW11 8LR;
    Tel: 0181 943 5331, Fax: 0181 943 4351

Land Infrared, Dronfield, Sheffield, S18 6DJ;   S
    Tel: 01246 417691, Fax: 01246 410585

Dr Bruno Lange (UK) Ltd, PO Box 417, Camberley,   A
    Surrey, GU15 3DU;
    Tel: 01276 677233, Fax: 01276 677307

Leeds & Northrup Ltd, Wharfdale Road, Tyseley,   CDLSW
    Birmingham, B11 2DJ;
    Tel: 0121 706 6171, Fax: 0121 706 1058

Lee-Integer Ltd, 1–3 Bowling Green Road, Kettering,   D
    Northants, NN15 7QW;
    Tel: 01536 511010, Fax: 01536 513653

Lindley Group Ltd, 385 Canal Road, Frizinghall,   EQW
    Bradford,West Yorks, BD2 1AX;
    Tel: 01274 530066, Fax: 01274 530084

Litre Meter Ltd, 50–53 Rabans Close, Rabans Lane   E
    Industrial Estate, Aylesbury, Bucks, HP19 3RS;
    Tel: 01296 436446, Fax: 01296 20341

Longlands College of Further Education Corp,   T
    Douglas Street, Middlesborough, Cleveland,
    TS4 2JW;
    Tel: 01642 248351, Fax: 01642 245313

Loughborough Projects Ltd, Swingbridge Road,   N
    Loughborough, Leics, LE11 0JB;
    Tel: 01509 262042, Fax: 01509 262517

Ludlow Sysco Ltd, Broadway, Market Lavington,   L
    Devizes, SN10 5RQ;
    Tel: 01380 818411, Fax: 01380 812733

M

Magnetrol International UK Ltd, Unit 1,   EG
    Regent Business Centre, Jubilee Road, Burgess Hill,
    West Sussex, RH15 9TL;
    Tel: 01444 871313, Fax: 01444 871317

Malvern Instruments Ltd, Spring Lane South,   BDLW
    Malvern, Worcs, WR14 1AT;
    Tel: 01684 892456, Fax: 01684 892789

Maywood Instruments Ltd, Rankine Road, Daneshill NP
    Industrial Estate, Basingstoke, Hants, RG24 8PP;
    Tel: 01256 57572, Fax: 01256 840937

MDA Scientific UK, 10 Boleyn Court, Tudor Road, R
    Manor Park, Runcorn, Cheshire, WA7 1SR;
    Tel: 01928 579224, Fax: 01928 579287

Measurement Group UK Ltd, Stroudly Road, CNST
    Basingstoke, Hants, RG24 8FW;
    Tel: 01256 462131, Fax: 01256 471441

Metax Ltd, Crowborough Hill, Crowborough, S
    East Sussex, TN6 2EB;
    Tel: 01892 669999, Fax: 01892 665117

Michell Instruments Ltd, Nuffield Close, DR
    Cambridge, CB4 1SS;
    Tel: 01223 424427, Fax: 01223 426557

Micronics Ltd, Unit 6, Slaidburn Crescent, Fylde Road E
    Trading Estate, Southport, Merseyside, PR9 9YF;
    Tel: 01704 232130, Fax: 01704 232133

Millipore UK Ltd, The Boulevard, Blackmoor Lane, LQWX
    Watford, Herts, WD1 8YW;
    Tel: 01923 816375, Fax: 01923 818297

Milltronics Ltd, Oak House, Bromyard Road, EGUV
    Worcester, WR2 5HP;
    Tel: 01905 748404, Fax: 01905 748430

Moisture Control & Measurement Ltd, Thorp Arch DPQ
    Trading Estate, Wetherby, West Yorks, LS23 7BJ;
    Tel: 01937 843927, Fax: 01937 842524

Monition Ltd, Bolsover Business Park, Station Road, HT
    Bolsover, Chesterfield, S44 6BD;
    Tel: 01246 825212, Fax: 01246 827999

Monitran Ltd, Monitor House, 33 Hazelmere Road, JSV
    Penn, Bucks, HP10 8AD;
    Tel: 01494 816569, Fax: 01494 812256

Moore Products Co (UK) Ltd, Copse Road, Lufton, CGPS
    Yeovil, Somerset, BA22 8RN;
    Tel: 01935 706262, Fax: 01935 706969

Muirhead Vactric Components Ltd, Oakfield Road, QW
    Penge, London, SE20 8EW;
    Tel: 0181 659 9090, Fax: 0181 659 9906

N

National Instruments UK Corp, 21 Kingfisher Court, C
    Hambridge Road, Newbury, Berks, RG14 5SJ;
    Tel: 01635 523545, Fax: 01635 523154

Neotronics Scientific Ltd, Western House,	DG
2 Cambridge Road, Stansted Mountfitchet,
Essex, CM24 8BZ;
Tel: 01279 814848, Fax: 01279 813926

Neural Technologies Ltd, 7a Lavant Street,	K
Petersfield, Hants, GU32 3EL;
Tel: 01730 260256, Fax: 01730 260466

Newson Gale Ltd, Van Gaver House,	Q
48/50 Bridgford Road, West Bridgford,
Nottingham, NG2 6AP;
Tel: 0115 982 2422, Fax: 0115 981 6460

Nicolet Instruments Ltd, Budbrooke Road,	CLW
Warwick, CV34 5XH;
Tel: 01926 494111, Fax: 01926 494452

NIKAT Associates, Orchard House, School Lane,	T
Mickle Trafford, Chester, CH2 4EF;
Tel: 01244 300668, Fax: 01244 300677

Nobel Systems Ltd, Murdock Road, Bedford,	NP
Beds, MK41 7PQ;
Tel: 01234 349241, Fax: 01234 325387

Novatron Ltd, Unit 34, Southwater Ind. Estate,	D
Horsham, West Sussex;
Tel: 01403 733012, Fax: 01403 733311

NSK-RHP UK Ltd, Mere Way, Ruddington Fields	V
Business Park, Ruddington, Notts, NG11 6JZ;
Tel: 0115 936 6600, Fax: 0115 936 6702

Nulectrohms Ltd, High Street, Meppershall,	S
Shefford, Beds, SG17 5LX;
Tel: 01462 813000, Fax: 01462 816807

O

Oatencourt Ltd, 12 Kingswood Court, Maidenhead,	S
Berks, SL6 1DD;
Tel: 01628 25342, Fax: 01628 25485

OEM-Automatic Ltd, Whiteacres, Cambridge Road,	AJNPS
Whetstone, Leics, LE8 6ZG;
Tel: 0116 284 9900, Fax: 0116 284 1721

Oilab Lubrication Ltd, Sutherland House,	LW
31 Sutherland Road, Wolverhampton, WV4 5AR;
Tel: 01902 334106, Fax: 01902 333010

Ometron Ltd, Worsley Bridge Road, London,	NUV
SE26 5BX;
Tel: 0181 461 5555, Fax: 0181 461 4628

Orbisphere Laboratories, Staveley Hall, Staveley,   BLR
   Chesterfield, Derbyshire, S43 3TW;
   Tel: 01246 280226, Fax: 01246 280227

Oriel Systems Ltd, Unit 2, 56 Pickwick Road,   C
   Corsham, Wilts, SN13 9BX;
   Tel: 01249 715307, Fax: 01249 715459

P

Paar Scientific, 594 Kingston Road, Raynes Park,   L
   London, SW20 8DN;
   Tel: 0181 540 8553, Fax: 01891 543 8727

Pacific Scientific Ltd, HIAC/ROYCO Division,   DLW
   11 Manor Courtyard, Hughenden Avenue,
   High Wycombe, Bucks, HP13 5RE;
   Tel: 01494 473232, Fax: 01494 472566

Pall Industrial Hydraulics Ltd, Europa House,   LW
   Havant Street, Portsmouth, PO1 3PD;
   Tel: 01705 303303, Fax: 01705 302506

Palmer Environmental Ltd,Ty Coch House,   F
   Llantarnam Park Way, Cwmbran, Gwent,
   NP44 3AW;
   Tel: 01633 489479, Fax: 01633 877857

Panametrics Ltd, Unit 2, Villiers Court, 40 Upper   DEGR
   Mulgrave Road, Cheam, Surrey, SM2 7AJ;
   Tel: 0181 643 5150, Fax: 0181 643 4225

Parker Hannifin plc, Filter Division, Peel Street,   LW
   Morley, Leeds, LS27 8EL;
   Tel: 0113 253 7921, Fax: 0113 252 3998

Partech Instruments Ltd, Eleven Doors, Charlestown,   LQW
   St. Austell, Cornwall, PL25 3NN;
   Tel: 01726 748856, Fax: 01726 68850

Particle Data Ltd, 2 Goose's Foot Estate, Kingstone,   LW
   Hereford, HR2 9HY;
   Tel: 01981 250479, Fax: 01981 251434

Particle Measuring Systems Europe Ltd,   LW
   39 Cornwell Business Park, Salthouse Road,
   Brackmills, Northampton, NN14 7EX;
   Tel: 01604 675111, Fax: 01604 675112

Particle Technology Ltd, PO Box 173, Foston,   LW
   Derbyshire, DE65 5NZ;
   Tel: 01283 520365, Fax: 01283 520412

Peek Measurement Ltd, Kingsworthy, Winchester,   EL
   Hants, SO23 7QA;
   Tel: 01962 883200, Fax: 01962 88553

Philips Weighing, York Street, Cambridge, CB1 2SH;   N
    Tel: 01223 374396, Fax: 01223 374330

PMV Instrumentation Ltd, 17 Somerford Business   P
    Park, Christchurch, BH23 3RU;
    Tel: 01202 480303, Fax: 01202 480808

Proaction Services, Unit 6, Whitelee, Mytholmeroyd,   W
    Halifax, West Yorks, HX7 5AE;
    Tel: 01422 884252, Fax: 01422 885913

Pruftechnik (UK) Ltd, Burton Road, 2 Streethay,   JSV
    Lichfield, Staffs, WS13 8LN;
    Tel: 01543 417722, Fax: 01543 417723

PSM Instrumentation Ltd, 19 Cuckfield Road,   EP
    Hurstpierpoint, West Sussex, BN6 9RP;
    Tel: 01273 835599, Fax: 01273 835951

Pump & Package Ltd, The Old Manse, 20 King Street,   EQ
    Desborough, Northants, NN14 2RD;
    Tel: 01536 762674, Fax: 01536 761973

Q

Quadrex Scientific, PO Box 79, Weybridge,   L
    Surrey, KT13 9RA;
    Tel: 01932 347648, Fax: 01932 336028

Quantitech Ltd, Unit 3, Old Wolverton Road,   DRV
    Old Wolverton, Milton Keynes, MK12 5NP;
    Tel: 01908 227722, Fax: 01908 227733

R

Radio Data Technology Ltd, 10 Tabor Place,   C
    Crittall Road, Witham, Essex, CM8 3YP;
    Tel: 01376 501255, Fax: 01376 501312

Raychem Ltd, Raychem House, Tangier Lane,   F
    High Street, Eton, Berks, SL4 6BD;
    Tel: 01753 856111, Fax: 01753 859130

RDP Electronics Ltd, Grove Street, Heath Town,   CJNPV
    Wolverhampton, WV10 0PY;
    Tel: 01902 457512, Fax: 01902 452000

Real Time Corrosion Management Ltd (CML),   B
    Rutherford House, Manchester Science Park,
    Manchester, M15 6SZ;
    Tel: 0161 232 2000, Fax: 0161 232 2001

Reutlinger (UK) Ltd, 88 Derby Road, Ripley,   V
    Derby, DE5 3HT;
    Tel: 01773 743567, Fax: 01773 746046

Rheology International, 6 Livingstone Circus,   L
  Gillingham, Kent, ME7 4HA;
  Tel: 01634 575661, Fax: 01634 280531

Rhopoint Systems, Rhopoint Ltd, Oxted, Surrey,   C
  RH8 9BB;
  Tel: 01883 722222, Fax: 01883 717245

Ringspann (UK) Ltd, 3 Napier Road, Bedford,   J
  MK41 0QS;
  Tel: 01234 342511, Fax: 01234 217322

Rivertrace Engineering Ltd, Unit 5, Astra Business   X
  Centre, Bonehurst Road, Salfords, Surrey, RH1 5TL;
  Tel: 01293 820810, Fax: 01293 820813

Rolls-Royce MatEval Ltd, 245/246 Europa   U
  Boulevard, Gemini Business Park, West Brook,
  Warrington, WA5 5TN;
  Tel: 01925 574868, Fax: 01925 444365

Roscow Technical Ltd (Sigrist), 1 Pembroke Avenue,   ADFL
  Waterbeach, Cambridge, CB5 9QR;
  Tel: 01223 860595, Fax: 01223 861819

Rosemount Ltd, Heath Place, Bognor Regis,   CEGPRS
  West Sussex, PO22 9SH;
  Tel: 01243 863121, Fax: 01243 867554

Rospen Industries Ltd, Unit 15, Oldends Lane   N
  Industrial Estate, Oldends Lane, Stonehouse,
  Glos, GL10 3RQ;
  Tel: 01453 825212, Fax: 01453 828279

Roth Scientific Co Ltd, Roth House,   LW
  12 Armstrong Mall, The Summit Centre,
  Southwood, Farnborough, Hants, GU14 0NR;
  Tel: 01252 513131, Fax: 01252 543609

Rotronic Instruments UK Ltd, Unit 2, Bluebird House, DS
  Povery Cross Road, Horley, Surrey, RH6 0BR;
  Tel: 01293 773330, Fax: 01293 774720

S

SAIC Ltd, 26 Craven Court, Stanhope Road,   CHV
  Camberley, Surrey, GU15 3BW;
  Tel: 01276 675511, Fax: 01276 676262

Scheme Engineering Ltd, Comet House, Calleva Park, EG
  Aldermaston, Berks, RG7 4QW;
  Tel: 017356 79151

Schenck Ltd, Station Approach, Bicester, Oxon,   V
  OX6 7BZ;
  Tel: 01869 321321, Fax: 01869 321111

Schlumberger Industries, Water and Industrial   CEG
    Measurement Division, Salmon Fields,
    Royton, Oldham, OL2 6BX;
    Tel: 0161 627 0333, Fax: 0161 627 0295

Scientific Atlanta Ltd, Home Park Estate, Kings   V
    Langley, Herts, WD4 8LZ;
    Tel: 019277 66133, Fax: 019277 69018

Scientific Computers Ltd, Premiere House, Betts Way,   K
    London Road, Crawley, West Sussex, RH10 2GB;
    Tel: 01293 403636, Fax: 01293 403641

Sensonics, Addison Road, Chesham, Bucks, HP5 2BD;   CTV
    Tel: 01494 774251, Fax: 01494 791094

Sensor Technology Ltd, PO Box 36, Banbury,   JLN
    Oxon, OX15 6JB;
    Tel: 01295 730746 (also fax)

Sensortek Ltd, PO Box 222, Bury St. Edmunds,   A
    Suffolk, IP28 6EE;
    Tel: 01284 728150, Fax: 01284 728155

Sensotec, Unit 2, 9 Vulcan Way, Sandhurst,   NPV
    Camberley, Surrey, GU17 8DB;
    Tel: 01252 877117, Fax: 01252 877699

Servomex plc, Jarvis Brook, Crowborough,   R
    East Sussex, TN6 3DU;
    Tel: 01892 652181, Fax: 01892 662253

SGS Redwood Ltd, Rosscliffe Road, Ellesmere Port,   LW
    South Wirral, Cheshire, L65 5AS;
    Tel: 0151 3554931, Fax: 0151 3563253

Shaw Moisture Meters, Rawson Road, Westgate,   D
    Bradford, West Yorks, BD1 3SQ;
    Tel: 01274 733582, Fax: 01274 370151

Sick – See Erwick Sick

Signal Instrument Co Ltd, Standards House,   DR
    Doman Road, Camberley, Surrey, GU15 3DW;
    Tel: 01276 682841, Fax: 01276 691302

Sigrist – see Roscow Technical

SKF Engineering Products Ltd, 2 Tanners Drive,   CLTUV
    Blakelands, Milton Keynes, MK14 5BN;
    Tel: 01908 618666, Fax: 01908 618717

Solartron Instruments, 124 Victoria Road,   BCV
    Farnborough, Hants, GU14 7PW;
    Tel: 01252 376666, Fax: 01252 544981

Solartron Transducers Ltd, 124 Victoria Road,   ELRV
    Farnborough, Hants, GU14 7PW;
    Tel: 01252 544433, Fax: 01252 547384

Solomat Ltd, Parsonage Road, Takeley,   DR
   Bishop's Stortford, Herts, CM22 6PU;
   Tel: 01279 871252, Fax: 01279 870377

Sonatest plc, Dickens Road, Milton Keynes,   U
   MK12 5QQ;
   Tel: 01908 316345, Fax: 01908 321323

Sony Broadcast & Professional UK, The Heights,   C
   Brooklands, Weybridge, Surrey, KT13 0XW;
   Tel: 01932 816000, Fax: 01932 817011

Southampton Institute of Higher Education,   T
   East Park Terrace, Southampton, SO9 4WW;
   Tel: 01703 319000, Fax: 01703 222259

Specac Ltd, River House, Lagoon Road, St. Mary   L
   Cray, Orpington, Kent, BR5 3QX;
   Tel: 01689 73134, Fax: 01689 78527

Spectro Analytical UK Ltd, Fountain House, Great   LW
   Cornbow, Halesowen, West Midlands, B63 3BL;
   Tel: 0121 550 8997, Fax: 0121 550 5165

SPM Instrument UK Ltd, PO Box 14, Bolholt Works,   FJNUVW
   Walshaw Road, Bury, Lancs, BL8 1PY;
   Tel: 0161 761 4837, Fax: 0161 797 5209

Status Instruments Ltd, Green Lane, Tewkesbury,   CDPS
   Glos, GL20 8HE;
   Tel: 01684 296818, Fax: 01684 293746

Staveley NDT Technologies, Inspection Instrument   LW
   Division, 19 Buckingham Avenue, Slough,
   Bucks, SL1 4QB;
   Tel: 01753 76216, Fax: 01753 821038

Steptech Instrument Services Ltd, Steptech House,   CDLMX
   Primrose Lane, Arlesey, Beds, SG15 6RD;
   Tel: 01462 733566, Fax: 01462 733909

Stewart Hughes Ltd, Chilworth Manor, Southampton,   V
   Hants, SO9 1XB;
   Tel: 01703 760222, Fax: 01703 768634

Stresswave Technology Ltd, Ashleigh House,   U
   Cromford Road, Wirksworth, Derby, DE4 4FR;
   Tel: 01629 825454, Fax: 01629 824844

Swansea Tribology Centre, Department of Mechanical   LTW
   Engineering, University College of Swansea,
   University of Wales, Singleton Park, Swansea, SA2 8PP;
   Tel: 01792 295534, Fax: 01792 295674

Sysco Analytics Ltd, Broadway, Market Lavington,   LX
   Devizes, Wilts, SN10 5RQ;
   Tel: 01380 818411, Fax: 01380 812733

Systech Instruments, Goodson Industrial Mews,   DR
　　Wellington Street, Thame, Oxon, OX9 3BX;
　　Tel: 01844 216838, Fax: 01844 217220

T

TEC Europe, Suite 8, Northwood House, Greenwood   V
　　Business Centre, Salford, Manchester, M5 4QH;
　　Tel: 0161 877 5773, Fax: 0161 877 5774

Techni Measure Ltd, Alexandra Buildings,   JNPV
　　59 Alcester Road, Studley, Warks, B80 7NJ;
　　Tel: 01527 854103, Fax: 01527 853267

Testo Ltd, Old Flour Mill, Queen Street, Emsworth,   CDJMSX
　　Hants, PO10 7BT;
　　Tel: 01243 377222, Fax: 01243 378013

Texcel Division CBS, Unit 8, Avebury Court,   EGPS
　　Mark Road, Hemel Hempstead, Herts, HP2 7TA;
　　Tel: 01442 231700, Fax: 01442 61918

Thermodata Components, Campton Road,   S
　　Meppershall, Shefford, Beds, SG17 5NN;
　　Tel: 01462 811757, Fax: 01462 811536

Towerite Ltd, Unit 3, Moulton Park Business Centre,   B
　　Redhouse Road, Moulton Park Industrial Estate,
　　Northampton, NN3 1AQ;
　　Tel: 01604 497521, Fax: 01604 497522

Transducer World (Europe) Ltd, PO Box 61,   NP
　　Aylesbury, Bucks, HP15 9YT;
　　Tel: 01296 437545, Fax: 01296 415149

TransInstruments Ltd, Lennox Road, Basingstoke,   GP
　　Hants, RG22 4AW;
　　Tel: 01256 20244, Fax: 01256 473680

Trevor M. Hunt, 50 Kingsholm Road, Westbury-on-   T
　　Trym, Bristol, BS10 5LH;
　　Tel: 0117 950 7194, Fax: 0117 950 7194

Tribometrics Inc, 2475 4th Street, Berkeley,   W
　　CA 94710, USA;
　　Tel: 001 510 540 1247, Fax: 001 510 527 7247

Trolex Ltd, Newby Road, Hazel Grove, Stockport,   CEJNPRSQV
　　Cheshire, SK7 5DY;
　　Tel: 0161 483 1435, Fax: 0161 483 5556

U

UCC International Ltd, PO Box 3, Thetford,   EGLPWX
　　Norfolk, IP24 3RT;
　　Tel: 01842 754251, Fax: 01842 753702

Unipro Ltd, 28 Campus Road, Listerhills Science Park, CH
    Bradford, West Yorks, BD7 1HR;
    Tel: 01274 740120, Fax: 01274 740136

United Air Specialists UK Ltd, Heathcote Way,   LQW
    Heathcote Industrial Estate, Warwick, CV34 6LY;
    Tel: 01926 311621, Fax: 01926 315986

V

Vaisala (UK) Ltd, Cambridge Science Park,   CDS
    Milton Road, Cambridge, CB4 4GH;
    Tel: 01223 420112, Fax: 01223 420988

VG Gas Analysis Systems Ltd, Unit 1, Aston Way,   R
    Middlewich, Cheshire, CW10 0HT;
    Tel: 01606 844731, Fax: 01606 845824

Vibro-meter Ltd, Bramhall Technology Park,   GNV
    Pepper Road, Hazel Grove, Cheshire, SK7 5BW;
    Tel: 0161 4830811, Fax: 0161 4832850

Vickers Systems Ltd, TEDECO Division, Larchwood   LW
    Avenue, Bedhampton, Havant, Hants, PO9 3QN;
    Tel: 01705 487260, Fax: 01705 492400

Vydas International Marketing, 1 Royal Parade,   N
    Hindhead, Surrey, GU26 6TD;
    Tel: 01428 606222, Fax: 01428 606676

W

Wearcheck Laboratories, Llandudno, North Wales,   LWX
    LL30 1SA;
    Tel: 01492 581811, Fax: 01492 585290

Wolfson Centre for Bulk Solids Handling   DER
    Technology, Unit 8, Block 1, Woolwich Dockyard
    Industrial Estate, Woolwich Church Street,
    Woolwich, London, SE18 5PQ;
    Tel: 0181 3318646, Fax: 0181 3318647

Wolfson Maintenance, Enterprise House,   TWV
    Manchester Science Park, Lloyd Street North,
    Manchester, M15 4EN;
    Tel: 0161 2263378, Fax: 0161 2263464

References

Anonymous (1995) *Water Services* (January), 15.

Copper Development Association (1995) Common Quality Problems and Best Practice Solutions, *Publication 111*, February 1995.

Data Cell (1994) In Proceedings of the Condition Monitoring Conference, University College of Swansea, 1994 (ed. M.H. Jones), Pineridge Press .

Enwald, P.A. and Liu, G.Q. (1994) In Proceedings of the Condition Monitoring Conference, University College of Swansea, 1994 (ed. M.H. Jones), Pineridge Press.

Er, M.J., *et al.* (1995) *Insight*, 37(1) (January), 31–5.

Fenner (1991) *Fenner Application Example* 430/91.

Garvey, R. (1994) In Proceedings of the Condition Monitoring Conference, University College of Swansea, 1994 (ed. M.H. Jones), Pineridge Press.

Hernu, M. (1992) Maintenance excellence – a comparative survey of leading European plants. *Maintenance*, 7(1) (March).

Holroyd, T.J. and Randall, N. (1994) In Proceedings of the Condition Monitoring Conference, University College of Swansea, 1994 (ed. M.H. Jones), Pineridge Press.

Hunt, T.M (1993) *Handbook of Wear Debris Analysis and Particle Detection in Liquids*, Chapman & Hall, London.

Mathur, A. (1994) In Proceedings of the 6th COMADEM Conference, (ed. B.K.N. Rao), Tata McGraw-Hill, New Delhi.

Pei, S. *et al.* (1995) *Insight*, 37(1) (January), 21–4.

Real Time Corrosion Management (undated) *CML Applications Note CML-LO8*.

Smith, E. (1992) *Laboratory Practice*, 41(12), 15–16.

Witt, K. *et al.* (1977) In Proceedings of the ICMES Conference, Paris (May).

Yates, M.A. (1994) In Proceedings of the Condition Monitoring Conference, University College of Swansea, 1994 (ed. M.H. Jones), Pineridge Press.

Further reading

This current book has introduced many ideas and techniques. It should have directed the reader to specific types of monitoring appropriate to his or her application. However, although the information is sufficient for decisions to be made, and in some cases to gain considerable understanding, there will possibly be the need to delve deeper. This bibliography presents a brief selection of other literature, some of which is of a more advanced nature. The books and papers are listed under headings relating to the type of monitoring.

BOOKS

Building and Engineering Services

ARMSTRONG (86), *Condition monitoring – an introduction to its application in building services*, J. Armstrong, BSIRA, 1986.

ARMSTRONG (88), *Low cost condition monitoring for engineering services*, J.H. Armstrong and P. Taylor, E. & F. Spon, 1988. ISBN 0419144501.

Fluid power

WATTON (92), *Condition monitoring and fault diagnosis in fluid power systems*, J. Watton, Ellis Horwood, 1992, 271 pages. ISBN 0131764055.

General monitoring

BOVING (90), *Non-destructive examination methods for condition monitoring*, K.G. Boving, NDE Handbook, Butterworth, 1990.

COLLACOTT (77), *Mechanical fault diagnosis and condition monitoring*, R.A. Collacott. Chapman & Hall, 1977. ISBN 0470990953.

DAVIES (96), *Handbook of condition monitoring*, A. Davies, Chapman & Hall, 1996.

HOPE (90), *Condition monitoring technology*, A.D. Hope, Longman, 1990. ISBN 0582023554.

NEALE (79), *Guide to the condition monitoring of machinery*, M.J. Neale, HMSO, 1979.

WILD (94), *Industrial sensors and applications for condition monitoring*, P. Wild, Mechanical Engineering Publishers, 1994, 127 pages. ISBN 0852989024.

Maintenance strategies

HIGGINS (88), *Maintenance engineering handbook*, 4th edition, L.R. Higgins, McGraw-Hill, 1988. ISBN 007028766X

HOLMBERG (91), *Operational reliability and systematic maintenance*, K. Holmberg and A. Folkeson (editors), Elsevier Applied Science, 1991. ISBN 1851666125.

LYONNET (91), *Maintenance planning*, P. Lyonnet, (1988, translated by J. Howlett), Chapman & Hall, 1991. ISBN 0412376806 (paperback).

MOUBRAY (91), *Reliability-centred maintenance*, J. Moubray, Butterworth–Heinemann, Oxford, 1991, 352 pages.

RAO, (96), *Handbook on condition monitoring and maintenance management in industries*, R.B.K.N. Rao, Elsevier Advanced Technology, 1996. ISBN 1856172341.

WILSON (84), *Maintenance managers guide to computer applications*, A. Wilson, DTI, 1984, 143 pages.

Management

DAVIES (90), *Management guide to condition monitoring in manufacturing*, A. Davies (editor), I. Prod. E. 1990.

KELLY & HARRIS (83), *Management of industrial maintenance*. Kelly and Harris, 1983, Butterworth IEE.

Compilation of several general papers and Conference Proceedings

BHRGroup (86–), *Condition Monitoring*, Conference Proceedings, 1986 (1st Brighton), 1988 (2nd London), 1990 (3rd London), 1992 (Stratford-upon-Avon, *Profitable Condition Monitoring*).

COMADEM (88–), Conference Proceedings, B.K.N. Rao (editor), 1988 (1st UK Birmingham), 1989 (1st International Birmingham), 1990 (2nd Uxbridge), 1991 (3rd Southampton), 1992 (4th Senlis, France), 1993 (5th Bristol), 1994 (6th New Delhi).

ERA (85), *Condition monitoring in hostile environments*, June 1985, BInstNDT Seminar Proceedings.

IMECHE(85[1]), *Vehicle condition monitoring and fault diagnosis*, March 1985, IMechE Conference.

IMECHE (85[2]), *Condition monitoring of machinery and plant*, June 1985, IMechE Conference.

IMECHE (90), *Machine condition monitoring*, 1990 IMechE Solid Mechanics and Machine Systems Groups Conference.

IMECHE (92), *Condition monitoring of building services systems*, April 1992, IMechE Conference.

McEWEN (91), *Condition monitoring*, J.R. McEwen, Elsevier. ISBN 1855980126 (papers compiled from BHR Group Conference).

SCI (83), *On-line monitoring of continuous process plants*, D.W. Butcher (editor), 1983, Society of Chemical Industry Conference, London, Ellis Horwood. ISBN 0853126836.

SWANSEA (84–), *Condition Monitoring*, Proceedings arranged at or from University College of Swansea, M.H. Jones (editor), 1984, 1987, 1991 (in Erding), 1994, Pineridge Press.

Teaching notes

IEE Distance Learning, PCM Course Pack.

Vibration

BROCH (80), *Mechanical vibration and shock measurements*, 2nd Edn, Brüel and Kjaer, 1980.

LYON (89), *Machinery vibration condition monitoring*, Butterworth, 1989.

Wear debris analysis and oil analysis

ANDERSON (91), *Wear particle atlas*, revised edition, Spectro Inc., 1991, 192 pages.

BRITISH COAL (84), *Wear particle atlas*, British Coal, 1984, 16 pages.

HUNT (93), *Handbook of wear debris analysis and particle detection in liquids*, T.M. Hunt, Chapman & Hall, 1993, 490 pages. ISBN 1851669620.

SWANSEA TRIBOLOGY CENTRE (90), *A guide to wear particle recognition – for use with the Rotary Particle Depositor (RPD)*, Swansea, 1990, 40 pages.

MAGAZINES AND JOURNALS

Condition Monitor – an International Newsletter. BHRGroup. Elsevier Science Publishers Ltd. Editor S. Barrett (Elsevier Advanced Technology, Oxford), ISSN 0268-8050. Reviews.

Condition Monitoring and Diagnostic Technology (1990–1992 only) BInstNDT. Editor Raj. Rao. ISSN 0957-7661. (Later articles included in the *Journal of the BInstNDT*.)

International Journal of Adaptive Control and Signal Processing. John Wiley & Son. Editor M.J. Gimble (Department of Electronic and Electric Engineering, University of Strathclyde, Glasgow) ISSN 0890-6327. Neural networks etc.

Index